T0130734

ATLAS OF THE
MAMMALS
OF GREAT BRITAIN AND
NORTHERN IRELAND

ATLAS OF THE MAMMALS
OF GREAT BRITAIN AND NORTHERN IRELAND

DEREK CRAWLEY, FRAZER COOMBER,
LAURA KUBASIEWICZ, COLIN HARROWER, PETER EVANS,
JAMES WAGGITT, BETHANY SMITH & FIONA MATHEWS

PELAGIC PUBLISHING

Compiled and edited by:

Derek Crawley[1]*, Frazer Coomber[1,2]*, Laura Kubasiewicz[1], Colin Harrower[3], Peter Evans[4,5], James Waggitt[4,5], Bethany Smith[1] and Fiona Mathews[1,2].

* Made equal contributions

1. The Mammal Society, 8 St John's Church Road, London E9 6EJ.
2. The University of Sussex, John Maynard Smith Building, Falmer BN1 9QG.
3. NERC Centre for Ecology and Hydrology, Crowmarsh Gifford, Wallingford OX19 8BB.
4. Sea Watch Foundation, Ewyn y Don, Bull Bay, Amlwch, Anglesey LL68 9SD.
5. School of Ocean Sciences, Bangor University, Menai Bridge, Anglesey LL59 5AB.

Species accounts written by:

Diana Bell, Sam Berry, Johnny Birks, Jenny Bryce, Simone Bullion, Ruairidh Campbell, Roisin Campbell-Palmer, Anna Champneys, Paul Chanin, Arnold Cooke, Frazer Coomber, Derek Crawley, Elizabeth Croose, Mike Dean, John Dutton, Peter Evans, John Flowerdew, Anita Glover, Martyn Gorman, John Gurnell, John Haddow, Daniel Hargreaves, Jochen Langbein, Penny Lewns, Fiona Mathews, Robbie McDonald, Tony Mitchell-Jones, Marion O'Neil, Josephine Pemberton, Rory Putman, Nigel Reeve, Hugh Rose, Henry Schofield, Dawn Scott, Jo Sharplin, Craig Shuttleworth, Peter Smith, Charles Smith-Jones, David Thompson, Roger Trout, Alastair Ward, Daniel Whitby and Ian White.

Published by Pelagic Publishing
PO Box 874
Exeter
EX3 9BR
UK

www.pelagicpublishing.com

Atlas of the Mammals of Great Britain and Northern Ireland

ISBN 978-1-78427-204-3   *Hardback*
ISBN 978-1-78427-205-0   *ePub*
ISBN 978-1-78427-206-7   *PDF*

© The Mammal Society 2020

The moral rights of the authors have been asserted.

All rights reserved. Apart from short excerpts for use in research or for reviews, no part of this document may be printed or reproduced, stored in a retrieval system, or transmitted in any form or by any means, electronic, mechanical, photocopying, recording, now known or hereafter invented or otherwise without prior permission from the publisher.

A CIP record for this book is available from the British Library

Cover photograph: Dan Rushton
Printed and bound in India by Replika Press Pvt. Ltd.

CONTENTS

FOREWORD   viii

ACKNOWLEDGEMENTS   x

INTRODUCTION   I

METHODS   3
    Data collection   3
    Distribution maps   5
    Record coverage across the UK   5

SPECIES ACCOUNTS   13
    Hedgehog (*Erinaceus europaeus*; Linnaeus, 1758)   14
    European mole (*Talpa europaea*; Linnaeus, 1758)   16
    Common shrew (*Sorex araneus*; Linnaeus, 1758)   18
    Pygmy shrew (*Sorex minutus*; Linnaeus, 1766)   20
    Water shrew (*Neomys fodiens*; Pennant, 1771)   22
    Lesser white-toothed shrew (*Crocidura suaveolens*; Pallas, 1811) 24
    European rabbit (*Oryctolagus cuniculus*; Linnaeus, 1758) 26
    Brown hare (*Lepus europaeus*; Pallas, 1778)   28
    Mountain hare (*Lepus timidus*; Linnaeus, 1758)   30
    Red squirrel (*Sciurus vulgaris*; Linnaeus, 1758)   32
    Grey squirrel (*Sciurus carolinensis*; Gmelin, 1788) 34
    Eurasian beaver (*Castor fiber*; Linnaeus, 1758)   36
    Hazel dormouse (*Muscardinus avellanarius*; Linnaeus, 1758) 38
    Edible dormouse (*Glis glis*; Brisson, 1762)   40
    Bank vole (*Myodes glareolus*; Schreber, 1780)   42
    Field vole (*Microtus agrestis*; Linnaeus, 1761)   44
    Common vole (*Microtus arvalis*; Pallas, 1778)   46
    Water vole (*Arvicola amphibius*; Linnaeus, 1758)   48
    Harvest mouse (*Micromys minutus*; Pallas, 1771)   50
    Wood mouse (*Apodemus sylvaticus*; Linnaeus, 1758) 52
    Yellow-necked mouse (*Apodemus flavicollis*; Melchior, 1834) 54
    House mouse (*Mus musculus*; Linnaeus, 1758)   56

Brown rat (*Rattus norvegicus*; Berkenhout, 1769)   58

Black rat (*Rattus rattus*; Linnaeus, 1758)   60

Wildcat (*Felis silvestris*; Miller, 1907)   62

Red fox (*Vulpes vulpes*; Linnaeus, 1758)   64

Badger (*Meles meles*; Linnaeus, 1758)   66

Otter (*Lutra lutra*; Linnaeus, 1758)   68

Pine marten (*Martes martes*; Linnaeus, 1758)   70

Stoat (*Mustela erminea*; Linnaeus, 1758)   72

Weasel (*Mustela nivalis*; Linnaeus, 1758)   74

Polecat (*Mustela putorius*; Linnaeus, 1758)   76

American mink (*Neovison vison*; Schreber, 1777)   78

Wild boar (*Sus scrofa*; Linnaeus, 1758)   80

Red deer (*Cervus elaphus*; Linnaeus, 1758)   82

Sika deer (*Cervus nippon*; Temminck, 1838)   84

Fallow deer (*Dama dama*; Linnaeus, 1758)   86

Roe deer (*Capreolus capreolus*; Linnaeus, 1758)   88

Chinese water deer (*Hydropotes inermis*; Swinhoe, 1870)   90

Reeves' muntjac deer (*Muntiacus reevesi*; Ogilby, 1839)   92

Greater horseshoe bat (*Rhinolophus ferrumequinum*; Schreber, 1774)   94

Lesser horseshoe bat (*Rhinolophus hipposideros*; Bechstein, 1800)   96

Alcathoe bat (*Myotis alcathoe*; von Helversen & Heller, 2001)   98

Whiskered bat (*Myotis mystacinus*; Kuhl, 1817)   100

Brandt's bat (*Myotis brandtii*; Eversmann, 1845)   102

Bechstein's bat (*Myotis bechsteinii*; Kuhl, 1817)   104

Daubenton's bat (*Myotis daubentonii*; Kuhl, 1817)   106

Greater mouse-eared bat (*Myotis myotis*; Borkhausen, 1797)   108

Natterer's bat (*Myotis nattereri*; Kuhl, 1817)   110

Serotine bat (*Eptesicus serotinus*; Schreber, 1774)   112

Leisler's bat (*Nyctalus leisleri*; Kuhl, 1817)   114

Noctule bat (*Nyctalus noctula*; Schreber, 1774)   116

Common pipistrelle bat (*Pipistrellus pipistrellus*; Schreber, 1774)   118

Soprano pipistrelle bat (*Pipistrellus pygmaeus*; Leach, 1825)   120

Nathusius' pipistrelle bat (*Pipistrellus nathusii*; Keyserling & Blasius, 1839)   122

Barbastelle bat (*Barbastella barbastellus*; Schreber, 1774)   124

Brown long-eared bat (*Plecotus auritus*; Linnaeus, 1758)   126

Grey long-eared bat (*Plecotus austriacus*; Fischer, 1829)   128

Grey seal (*Halichoerus grypus*; Fabricius, 1791)   130

Harbour seal (*Phoca vitulina*; Linnaeus, 1758)   132

North Atlantic right whale (*Eubalaena glacialis*; Müller, 1776)   134

Bowhead whale (*Balaena mysticetus*; Linnaeus, 1758)   136

Humpback whale (*Megaptera novaeangliae*; Borowski, 1781)   138

Blue whale (*Balaenoptera musculus*; Linnaeus, 1758)   140

Fin whale (*Balaenoptera physalus*; Linnaeus, 1758)   142

Sei whale (*Balaenoptera borealis*; Lesson, 1828)   144

Minke whale (*Balaenoptera acutorostrata*; Lacepede, 1804)   146

Northern bottlenose whale (*Hyperoodon ampullatus*; Forster, 1770) 148

Cuvier's beaked whale (*Ziphius cavirostris*; Cuvier, 1823) 150

Sowerby's beaked whale (*Mesoplodon bidens*; Sowerby, 1804) 152

Pygmy sperm whale (*Kogia breviceps*; Blainville, 1838) 154

Dwarf sperm whale (*Kogia sima*; Owen, 1866) 156

Sperm whale (*Physeter macrocephalus*; Linnaeus, 1758) 158

Beluga (*Delphinapterus leucas*; Pallas, 1776) 160

Killer whale or Orca (*Orcinus orca*; Linnaeus, 1758) 162

False killer whale (*Pseudorca crassidens*; Owen, 1846) 164

Long-finned pilot whale (*Globicephala melas*; Traill, 1809) 166

Risso's dolphin (*Grampus griseus*; Cuvier, 1812) 168

Atlantic white-sided dolphin (*Lagenorhynchus acutus*; Gray, 1828) 170

White-beaked dolphin (*Lagenorhynchus albirostris*; Gray, 1846) 172

Common dolphin (*Delphinus delphis*; Linnaeus, 1758) 174

Striped dolphin (*Stenella coeruleoalba*; Meyen, 1833) 176

Bottlenose dolphin (*Tursiops truncatus*; Montagu, 1821) 178

Harbour porpoise (*Phocoena phocoena*; Linnaeus, 1758) 180

CETACEANS KNOWN IN BRITAIN AND IRELAND ONLY FROM STRANDINGS 182

Blainville's beaked whale (*Mesoplodon densirostris*; Blainville, 1817) 182

Gervais' beaked whale (*Mesoplodon europaeus*; Gervais, 1855) 182

True's beaked whale (*Mesoplodon mirus*; True, 1913) 183

Narwhal (*Monodon monoceros*; Linnaeus, 1758) 183

Melon-headed whale (*Peponocephala electra*; Gray, 1846) 184

Fraser's dolphin (*Lagenodelphis hosei*; Fraser, 1956) 184

VAGRANT SPECIES AND THOSE WITHOUT ESTABLISHED POPULATIONS IN THE UK 185

Raccoon (*Procyon lotor*; Linnaeus, 1758) 185

Red-necked wallaby (*Macropus rufogriseus*; Desmarest, 1817) 185

Reindeer (*Rangifer tarandus*; Linnaeus, 1758) 186

Pond bat (*Myotis dasycneme*; Boie, 1825) 186

Geoffroy's bat (*Myotis emarginatus*; Geoffroy, 1806) 186

Northern bat (*Eptesicus nilssonii*; Keyserling and Blasius, 1839) 187

Parti-coloured bat (*Vespertilio murinus*; Linnaeus, 1758) 187

Kuhl's pipistrelle bat (*Pipistrellus kuhlii*; Kuhl, 1817) 188

Savi's pipistrelle bat (*Hypsugo savii*; Bonaparte, 1837) 188

Bearded seal (*Erignathus barbatus*; Erxleben, 1777) 189

Harp seal (*Pagophilus groenlandicus*; Erxleben, 1777) 189

Hooded seal (*Cystophora cristata*; Erxleben, 1777) 189

Ringed seal (*Pusa hispida*; Schreber, 1775) 190

Walrus (*Odobenus rosmarus*; Linnaeus, 1758) 190

FERAL COLONIES AND POPULATIONS 191

FOREWORD

While the maps in this Atlas are fascinating in their own right, they are also an important resource for the conservation and management of mammal populations. Fulfilling the Society's role in advocating science-led mammal conservation, they underlie the generalised distribution maps provided in the *Guide to the Population and Conservation Status of Britain's Mammals* (Mathews *et al.*, 2018). In this Atlas, information on reported occupancy is shown in higher spatial resolution, and the maps also document changes in recorded distribution since 1960. They also provide an essential reference for ecological consultants wishing to know which species of protected mammals could potentially be affected by a development in a particular area.

This is the latest in a series of publications based on the systematic collection of distribution records by the Mammal Society and its members extending back to the 1960s. Provisional Atlases were published in 1971, 1978 and 1984, while the first *Atlas of Mammals in Britain*, compiled and edited by Henry Arnold, was published by the Institute of Terrestrial Ecology and the Joint Nature Conservation Committee in 1993. In addition to Henry, who also edited the second and third provisional Atlases, and to Gordon Corbet, editor of the first of these, a small army of people must be thanked for their contributions. This includes you, the reader; if you have ever submitted a mammal record, whether directly, through one of our partners, or via the Society's apps – *Mammal Tracker* or *Mammal Mapper* – I would like to thank you on behalf of the Society.

Over the past five years a small group of people have co-ordinated collectors of records, cajoled contributors and tried to coerce computers into providing suitable maps for this Atlas. Derek Crawley deserves a special mention and my undying gratitude for taking on the co-ordinating role, following the untimely death of Derek Yalden, my predecessor as President. Fiona Mathews, Laura Kubasiewicz, Frazer Coomber, Peter Evans, James Waggitt, Colin Harrower, Bethany Smith, Richard Shore, David Roy and Martin Harvey worked with Derek to turn a collection of distribution records into an informative atlas for which we are very grateful. The Society is also deeply indebted to the mammal groups, mammal recorders, record verifiers and authors of the species accounts who all played their part in creating this volume.

There is one person, however, whose contribution overshadows all of these: Derek Yalden was an inspirational mammalogist whose boundless enthusiasm and extensive knowledge infected generations of naturalists with the mammal-recording bug. Derek's unexpected death in 2013 was a great blow to the Society. In an obituary for Derek (*Mammal Notes*, no. 166, 2013) Pat Morris recalls that he and Derek, who

met while still at school, 'took up the challenge to fill in blank squares on [the] distribution maps' produced for the London Natural History Society's mammal recorder.
This habit persisted throughout Derek's life and was particularly, but by no means
exclusively, focused on the Peak District where he studied the distribution and
ecology of mountain hares, red-necked wallabies and even birds such as the golden
plover and common sandpiper. Derek was a member of the Society for 50 years and
filled many significant roles. He was editor of *Mammal Review* for 32 years; wrote or
contributed many technical publications, including all four editions of the *Handbook
of British Mammals* (co-editing the last of these, an 800-page blockbuster published
by the Society in 2008); and was President for 16 years, a period equalled only by
the first Society President, the fourth Earl of Cranbrook, John Gathorne-Hardy.

During his presidency, Derek proposed that the Society should produce a
Mammal Atlas for the twenty-first century and was the driving force behind its
development. This volume is a testament to his vision, his enthusiasm and his great
talent for involving other people in studying mammals.

We owe him a great deal and we still miss him.

Paul Chanin, June 2018

ACKNOWLEDGEMENTS

We would like to thank the many volunteer surveyors and ecological consultants who contributed the data used to produce the Atlas distribution maps. We are also grateful for the support of local Biological Records Centres and other organisations in sharing their datasets: our task would have been much harder without the tremendous contribution of county recorders and verifiers. Tom Hunt at the Association of Local Records Centres provided valuable assistance in ensuring the smooth transition of data, and we are also grateful to the National Biodiversity Network Trust. Specific records were supplied by Baseline Ecology, the Bat Conservation Trust, Bedfordshire and Luton Biodiversity Recording and Monitoring Centre, Biodiversity Gatwick Project, Biodiversity Information Service for Powys and Brecon Beacons National Park, Biological Records Centre, Boat of Garten Wildlife Group, Bristol Regional Environmental Records Centre, British Deer Society, British Trust for Ornithology, Buckinghamshire and Milton Keynes Environmental Records Centre, Cambridgeshire & Peterborough Environmental Records Centre, Canals and Rivers Trust, Central Scotland Green Network Trust, Cofnod (North Wales Environmental Information Service), Cumbria Biodiversity Data Centre, Derbyshire Biological Records Centre, Derbyshire Mammal Atlas, Derbyshire Mammal Group, Devon Biodiversity Records Centre, Doncaster MBC Biological Records Centre, Dorset Environmental Records Centre, Dorset Mammal Group, Dr Francis Rose Field Notebook Project, Dumfries and Galloway Environmental Resources Centre, East Ayrshire Countryside Ranger Service, EcoRecord, Environment Agency, Environmental Records Centre for Cornwall and the Isles of Scilly, Environmental Records Information Centre North East, Essex Wildlife Trust, Fife Nature Records Centre, Focus Ecology Ltd, Furesfen Ecological Consultancy, Glasgow Museums BRC, Gloucestershire Centre for Environmental Records, Greater Lincolnshire Nature Partnership, Greater Manchester Ecology Unit, Greenspace Information for Greater London (GiGL), Hampshire Biodiversity Information Centre, Herefordshire Biological Records Centre, Hertfordshire Natural History Society, Herts Environmental Records Centre, Highland Biological Recording Group, Humber Environmental Data Centre, Inner Forth Nature Counts, iRecord, Isle of Wight Local Records Centre, iSpot, IW Council Parks and Countryside Section, John Muir Trust, Kent & Medway Biological Records Centre, Kent Biological Records Centre, Lancashire Environment Record Network, Leicestershire and Rutland Environmental Records Centre, Lorn Natural History Group, Lothian and Borders Mammal Group, Lymington Naturalists, Mammals of Suffolk, MaWSE project, Merseyside BioBank, Ministry of Justice, MKA Ecology Ltd, National Trust, National Trust for Scotland, Natural Resources Wales, Nonsuch Watch, Norfolk

Biodiversity Information Service, North & East Yorkshire Ecological Data Centre, North Ayrshire Countryside Ranger Service, North East Scotland Biological Records Centre, Northamptonshire Biodiversity Records Centre, Nottinghamshire Mammal Database, Outer Hebrides Biological Recording Project, People's Trust for Endangered Species, Phlorum Consultancy, PJC Consultancy, Powys and BBNP Biodiversity Information Service, Preston Montford Field Studies Council Centre, Riverbank Wildlife Area, Rotherham Biological Records Centre, Royal Horticultural Society, RPS, RSPCA, SCC Open Space, Scottish Natural Heritage, Scottish Wildlife Trust, Sheffield Biological Records Centre, Shetland Biological Records Centre, Shire Group of Internal Drainage Boards, Shropshire Ecological Data Network, Somerset Environmental Records Centre, Sorby Mammal Group, Sorby Natural History Society, South East Wales Biodiversity Records Centre, St Helens Wildlife Recording Group, Staffordshire Ecological Record, Suffolk Biological Records Centre, Surrey Biodiversity Information Centre, Surrey Biological Records Centre, Surrey Dormouse Group, Surrey Mammal Group, Sussex Biodiversity Record Centre, Sustrans, Tawny Croft Wildlife Consultants, Thames Valley Environmental Records Centre, The Biodiversity Information System for Cheshire, Halton, Warrington and the Wirral, The Ecology Consultancy, The Magnificent Science Company Limited, Unsted Wildlife Monitoring, Vincent Wildlife Trust, Warwickshire Biological Records Centre, West Wales Biodiversity Information Centre, West Yorkshire Ecology Service, Wildlife Information Centre, WildWatch Project Records, Wiltshire and Swindon Biological Records Centre, Worcestershire Biological Records Centre, Yorkshire Naturalists' Union, Yorkshire Wildlife Trust, Declan Barraclough, Johnny Birks, A. Blunden, Ian Bond, I. Boyd, Helen Butler, Paul Chanin, Robert and Val Clinging, Derek Crawley, John Dobson, John Durkin, Dave and Joyce Earl, Peter Follett, Liz Halliwell, Annie Haycock, Zoe Haysted, Harriet James, David Jardine, Jenny Jones, Rosy Jones, G. Knass, Robert Lamb, Penny Lewns, Steve Lonsdale, Anne Marston, Chris Matcham, Brenda Mayles, Eve Mulholland, Ed Pooley, C.R. Pope, Brian Ribbands, Phil Richardson, Kirstie and Calum Ross, Graham Scholey, Paul Seligman, Rob Spencer, Stephanie Tyler, Tracy Underwood, Michael Walker, Debbie Wallace, Ken Walton and Antony Witts.

The data sources for cetacean records are shown in Table 1. These records were collated as part of the NERC/DEFRA funded Marine Ecosystems Research Programme (NE/L003279/1). We would especially like to thank: Joana Andrade, Mick Baines, Oliver Boisseau, Gareth Bradbury, Tom Brereton, Kees Camphuysen, Jose Martinez-Cedeira, Carol Cooper, Jan Durinck, Tom Felce, Isabel Garcia-Baron, Stefan Garthe, Steve Geelhoed, Anita Gilles, Martin Goodall, John Goold, Jan Haelters, Sally Hamilton, Phil Hammond, Lauren Hartny-Mills, Nicola Hodgins, John Houghton, Kathy James, Mark Jessopp, Nia Jones, Ailbhe Kavanagh, Mardik Leopold, Katrin Lohrengel, Maite Louzao, Colin MacLeod, Oliver O'Cadhla, Sarah Perry, Graham Pierce, Vincent Ridoux, Kevin Robinson, Camilo Saavedra, Begoña Santos, Richard Shucksmith, Henrik Skøv, Eric Stienen, Signe Sveegaard, Paul Thompson, Nicolas Vanermen, Dave Wall, Andy Webb and Jared Wilson.

We would also like to thank all the mammal record verifiers who have helped to ensure that the data are of the highest quality: Lorcan Adrain, Debbie Alston, Simone Bullion, Helen Butler, Rebecca Collins, Nathalie Cossa, Derek Crawley, John Ellis, Dan Forman, Claudia Gebhardt, Penny Green, Richard Grogan, Gareth Harris, Annie Haycock, Gary Hedges, Mark Hows, Barry Ingram, Jenny Jones, Jim Jones, Dave Kilbey, Ellie Knott, Emma Koblizek, Richard Lawrence, Ellen Lee, Sorcha Lewis, Declan Looney, John Mackintosh, Chris Manning, John Martin, Jacqueline

Merrick, Tony Mitchell-Jones, Phillip Morgan, Helen O'Brien, Gareth Parry, Jools Partridge, Katherine Pinnock, Jonathan Pounder, John Ray, Brian Ribbands, Andy Riches, Angela Ross, Andy Rothwell, David Roy, Ro Scott, Graham Smith, Tamasine Stretton, Sarah Underwood, Lisa Wade, Michael Walker, Richard Webb, David Williams, Graeme Wilson and John Young. Liam Lysaght has been helpful with ensuring co-ordination with the Atlas of Mammals of Ireland, and Johnny Birks, as then chair of the Mammal Society, was integral to setting up the current Atlas project.

Thanks are owed to the mammal experts who have written the species accounts, and their names are presented with each individual account. In addition, we would like to thank all the photographers who have submitted photographs to the Mammal Society, including entrants to our amateur Mammal Photographer of the Year competition.

The editors would like to apologise for any oversight in our thanks. A huge number of people have been involved with this project, submitting, collecting, preparing and verifying records; writing text; and by giving general support. Thank you all.

INTRODUCTION

The aim of the *Atlas of Mammals of Great Britain and Northern Ireland* is to increase understanding of the distribution of mammals, and to document changes over time since the last *Atlas* (1960–92; Arnold, 1993). The current project began as the Mammal Watch South East (MaWSE). Part-funded by the Heritage Lottery Fund (HLF), MaWSE created the *Mammal Tracker* app, which provided an easy inter-face for the public to submit records from across the UK. This local project then expanded to encompass the whole of the UK, the Channel Islands and the Isle of Man, and, with the assistance of the Sea Watch Foundation, was extended to include cetaceans.

The species accounts are ordered as terrestrial mammals (including seals), cetaceans, cetacean species known only from strandings, vagrant species and feral populations. The terrestrial mammals are arranged in the same order as the *Review of the Population and Conservation Status of British Mammals* (Mathews *et al.*, 2018) followed by the seal species, which were not included in that review. The cetacean species are organised according to taxonomic similarities. The common names used in this Atlas are those used in the *Review of the Population and Conservation Status of British Mammals*, or, for seals and cetaceans, those in the Mammal Society's online species hub. The *Atlas*, unlike the *Review*, includes the Channel Islands. The Orkney and Guernsey vole are therefore presented in a single species report as they are both the common vole *Microtus arvalis*. Island subspecies, such as the Skomer vole (a subspecies of the bank vole *Myodes glareolus*) and the Irish hare (a subspecies of the mountain hare *Lepus timidus*), are not presented individually.

Each species account, with the exception of the vagrant species, contains infor-mation on the species distribution, ecology and identification. This is accompanied by a distribution map, an illustration of the seasonal distribution of occurrence records, and a photograph intended to aid identification. For further information on the status of terrestrial mammals in Britain, readers are referred to the *Review of the Population and Conservation Status of British Mammals* and the Regional Red List for mammals that has been compiled by the Society for the Statutory Nature Conservation Organisations according to International Union for the Conservation of Nature (IUCN) standards. For cetaceans, the species accounts include sections on distribution, ecology and identification. These have been adapted for this *Atlas* from larger accounts written for the book *European Whales, Dolphins and Porpoises* to commemorate the 25th anniversary of the United Nations Environment Programme's regional agreement, ASCOBANS (Evans, 2019).

The photographs were sourced from the Mammal Society's catalogue of photo-graphs, species experts and various appeals. Many were submitted by members of the public as part of the Mammal Society's Mammal Photographer of the Year competition. Wherever possible, the *Atlas* uses photographs taken of wild animals within the UK and its Crown dependencies. Selection of photographs was based on the display of key identification features rather than aesthetic appeal. The Mammal Society is extremely grateful to all the photographers, and their names appear adjacent to the relevant image.

BIBLIOGRAPHY

ARNOLD, H.R. 1993. *Atlas of Mammals in Britain*. ITE Research Publication No. 6. London: Joint Nature Conservation Committee and Institute of Terrestrial Ecology, HMSO.

EVANS, P.G.H. 2019. *European Whales, Dolphins and Porpoises. Marine Mammal Conservation in Practice*. London and New York: Elsevier/Academic Press.

MAMMAL SOCIETY. 2018. Full species list. Accessed at http://www.mammal.org.uk/species-hub/uk-mammal-list/ (30 July 2019).

MATHEWS, F., KUBASIEWICZ, L.M., GURNELL, J., HARROWER, C.A., McDONALD, R.A. & SHORE, R.F. 2018. *A Review of the Population and Conservation Status of British Mammals. A Report by the Mammal Society Under Contract to Natural England, Natural Resources Wales and Scottish Natural Heritage*. Peterborough: Natural England.

METHODS

DATA COLLECTION

British mammal occurrence data at 10 km resolution or higher were gathered from the following sources: National Biodiversity Network (NBN) gateway (NBN Atlas), local biological record centres, local and national monitoring schemes, iRecord, Natural England, the Mammal Society, Sea Watch Foundation, the *Mammal Tracker* app, and records provided by individuals. A full list of data providers is supplied in the acknowledgements. Validation and verification was conducted by the Mammal Society's appointed verifiers, using protocols drawn up by Derek Crawley with help from the Biological Records Centre (BRC). Unverified mammal records derived from organisations that did not have mammal surveying as their primary objective were only used for those species that were unlikely to be misidentified, namely moles *Talpa europaea*, rabbits *Oryctolagus cuniculus*, badgers *Meles meles*, foxes *Vulpes vulpes* and hedgehogs *Erinaceus europaeus*. Most of these unverified records were supplied by the British Trust for Ornithology.

Thirty-six species of cetaceans have been recorded in European seas, 30 of which have been recorded at one time or another in the waters around Britain and Ireland. Species accounts are given for all 30 species, although distribution maps and full-page accounts are included only for those species recorded from live sightings. Six species have only been recorded from strandings. In Britain, these include Gervais' beaked whale *Mesoplodon europaeus*, Blainville's beaked whale *Mesoplodon densirostris*, narwhal *Monodon monoceros*, melon-headed whale *Peponocephala electra* and Fraser's dolphin *Lagenodelphis hosei*; in Ireland, True's beaked whale *Mesoplodon mirus* and Gervais' beaked whale are both known only from strandings. True's beaked whale is the only species recorded in Ireland but not in Britain.

The distribution maps presented here are compiled from verified sightings between 1980 and the present, held within the Sea Watch Foundation's national database and the NERC/Defra-funded Marine Ecosystem Research Programme (see Table 1 for details). These records are derived from observations made from the coast, from vessels, and from aerial surveys (both dedicated and opportunistic). Together, they total more than 200,000 records of live sightings. Strandings have not been included on the maps, although some live sightings near the coast were followed by the animal stranding.

Occurrence data are shown in this Atlas for two specific time periods: a historic period (1960–92; for cetaceans, 1980–99) and the current time period (2000–16). The historic period was chosen to coincide with the time frame of the 1993 mammal atlas (Arnold, 1993; hereafter referred to as the Arnold Atlas), to facilitate comparisons

TABLE 1   Sources of cetacean sightings data

Data source	No. of records	Data source	No. of records
Sea Watch Foundation	122,139	SAMM Survey	835
JNCC European Seabirds at Sea	14,527	Manx Whale & Dolphin Group	795
Dutch Seabird Group	10,447	Irish National Parks & Wildlife Service	701
The University of Veterinary Medicine Hannover	8,871	NORCET Surveys	669
WWT Consulting	6,796	MARINELife	616
JNCC MMO data	5,310	HiDef	449
Hebridean Whale & Dolphin Trust	4,988	CODA Survey	321
Cardigan Bay Marine Wildlife Centre	3,722	Aarhus University	227
Irish Whale & Dolphin Group	3,184	Marine Awareness North Wales	203
Institute for Marine Resources and Ecosystem Studies	3,015	Mardik Leopold	102
ORCA	2,194	Cornwall Wildlife Trust	85
Cetacean Research & Rescue Unit	2,192	SIAR Survey	80
SCANS-2 Survey	2,031	Coordinadora para o Estudo dos Mamíferos Mariños	65
Royal Belgium Institute for Natural Sciences	1,371	University of Swansea	58
Whale & Dolphin Conservation	1,240	RWE nPower	52
SCANS-1 Survey	1,198	PELGAS Survey	26
Bundesamt für Naturschutz	1,191	APEM	19
Instituut voor Natuur- en Bosonderzoek	928	ATLANCET Survey	1
Marine Conservation Research	911	JUVENA Survey	1
University of Aberdeen	911	**TOTAL**	**202,471**

between the two atlases. It should be noted that the information displayed in the current publication does not exactly replicate that shown for the same period in the Arnold Atlas. Ongoing updates to, and cleaning of, the dataset mean that there are additional data points that were missing from the Arnold Atlas; and conversely, some of the records shown in the Arnold Atlas have been removed or relocated.

Cetaceans were only sparsely and sporadically recorded before 1980, though records have become relatively common subsequently. Given the scarcity of data prior to the 1980s, the relative abundance of data in the 1990s, and the fact that the Arnold Atlas excluded cetaceans, there was little benefit from aligning the historical time period with the Arnold Atlas. Therefore, for cetaceans, the historical time period begins when recording became more common (1980) and ends at 1999, i.e. immediately before the current time period.

The current period for the Atlas is considered to be from 2000 to 2016 for both terrestrial mammals and cetaceans. The dates were set so that available data from published local county atlases could be used. However, for a few species that have experienced rapid changes in distribution – red squirrel *Sciurus vulgaris*, grey squirrel *Sciurus carolinensis* and water vole *Arvicola amphibius* – shorter time periods have been used. The current period is set at 2005–16 for the water vole, and 2010–16 for both squirrel species.

DISTRIBUTION MAPS

The distribution maps were produced from the collated occurrence data by Colin Harrower (for terrestrial species) and James Waggitt (cetaceans). Terrestrial mammal and seal maps were created using the 'BRCmap' package and the cetacean maps were produced using 'raster' (Hijmans, 2013) and 'maptools' (Bivand & Lewin-Koh, 2015) packages, within the analytical framework R (R Core Team, 2017). The data for both terrestrial mammals and cetaceans are mapped at a 10 km square resolution. The terrestrial species used the Ordnance Survey of Great Britain and Ordnance Survey of Ireland 10 km resolution grid squares. From here on, both marine and terrestrial grids will be referred to as hectads. For each hectad, a symbol is plotted to indicate whether a species was present in that hectad during the historical period, the current period, or both periods.

There may be some discrepancies between the Atlas maps and the known distributions of some species. This is because the maps were created using the available records for each species, and these records are not always complete. In some areas, for example the Scottish Highlands and Islands, the numbers of records are low (Figure 1). Lower recording effort may create artefactual gaps in distributions, and means that populations may not be well mapped on islands. In addition, data-sharing agreements mean that some records might not have been accessible. There is also the possibility of false presence records if the species can easily be misidentified, for instance stoat Mustela erminea and weasel Mustela nivalis, or bank and field voles Microtus agrestis. Although most errors of this kind have been removed during the data-cleaning process, it is very difficult to identify false positives where they lie within the known geographical range of the species and occur at an appropriate time of year.

It must also be stressed that a positive hectad represents a record, not necessarily an established breeding population. Therefore, the distribution maps display the true distribution from all verified records. As with all distribution atlases, the maps will only be as good as the available records from which they were created. So, many will represent the actual species distribution as well as, to some extent, the distribution of the observers.

Since cetaceans are very mobile and do not observe political boundaries, we have taken more natural areas and included records from across the North Sea and within the shelf seas surrounding Britain and Ireland. Figure 2 shows the boundaries of the area for which cetacean records have been presented and also includes the UK's Exclusive Economic Zone to represent the extent of British waters.

The seasonal distribution, or phenology, of records for each species is presented alongside the distribution maps. These phenology plots show a breakdown of the number of records per month to highlight the times of the year in which recordings for each species tend to be made. The phenology plots only include data from the current atlas period, 2000–16, and are only displayed for species with ten or more records.

RECORD COVERAGE ACROSS THE UK

Terrestrial mammal records from the current atlas period (2000–16) were collected from across the UK, the Channel Islands and the Isle of Man (Figure 1). In this Atlas, the terms 'UK' and 'Great Britain' are used as geographical terms that encompass the Crown dependencies, rather than political boundaries, and 'Ireland' refers to the island of Ireland. There was at least one record for most hectads including many islands, such as the Northern Isles, the Hebrides, the Isles of Scilly and the Isle of Wight. The number of records for each hectad ranged dramatically from one to over 2,000, and the distribution of records per hectad revealed distinct areas of low and

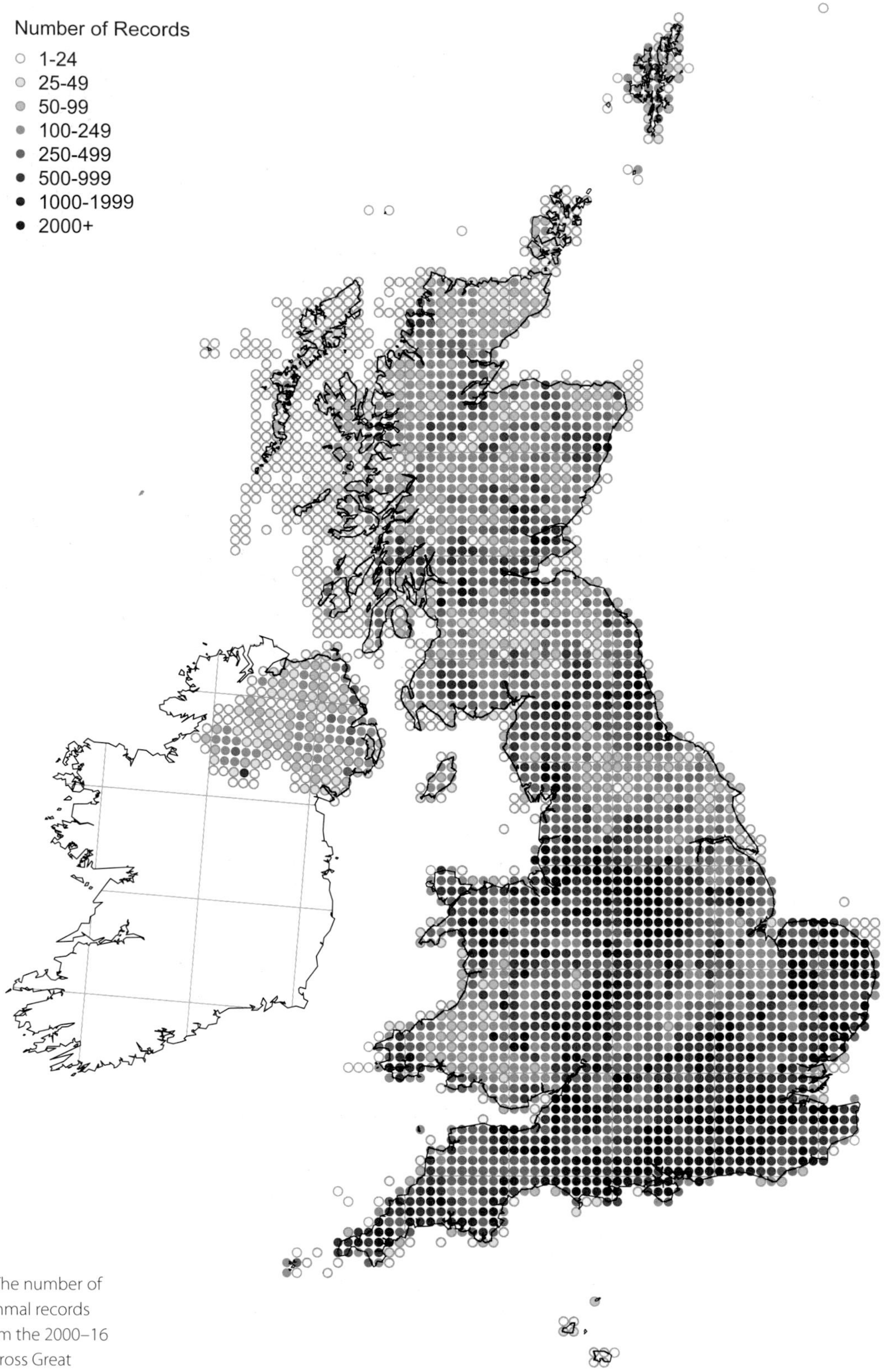

Number of Records
- ○ 1-24
- ◔ 25-49
- ◑ 50-99
- ◕ 100-249
- ● 250-499
- ● 500-999
- ● 1000-1999
- ● 2000+

FIGURE 1   The number of terrestrial mammal records per hectad from the 2000–16 atlas period across Great Britain and Northern Ireland.

FIGURE 2 Map of the area for which cetacean records have been reported. The UK's Exclusive Economic Zone has also been included to represent the extent of British waters.

high numbers of records. Areas with high numbers of records per hectad are located in the following regions: south of the line from the Bristol Channel to the Wash; Nottinghamshire; Derbyshire; Cumbria; and Northumberland. Areas of low numbers of records include the following: Pembrokeshire and Glamorgan; East Riding of Yorkshire; Northern Ireland; southern Scotland; and north-west Scotland (Figure 1).

Every species of terrestrial mammal covered in this Atlas, with the exception of the raccoon *Procyon lotor*, soprano pipistrelle bat *Pipistrellus pygmaeus* and the Alcathoe bat *Myotis alcathoe* (the latter two only being identified recently), has records from 2000 to 2016 and was present in at least one hectad in both the 2000–16 and the 1960–92 time periods. There were nearly 300,000 mammal occurrence records from the historical time period. This increased dramatically (by 435%) to just over 1.3 million records for the current period, despite the time period under consideration being shorter (17 years compared with 33 years), highlighting the increased recording of mammals between the two periods. All species except the common vole, wildcat *Felis silvestris* and house mouse *Mus musculus* had more records in the current than the historical time period (Table 2). Owing to very small numbers of records, black rat *Rattus rattus* and greater mouse-eared bat *Myotis myotis* are excluded from time-trend assessments.

There have been some noticeable changes in the distribution of some species: of the 60 resident terrestrial mammals, 13 have shown a decline in the number of occupied hectads. Declines of >30% were seen for the wildcat, red squirrel, harvest mouse *Micromys minutus* and house mouse. It is also apparent that some mustelids,

TABLE 2   The number of terrestrial mammal occurrence records and the number of occupied hectads. The final column shows the change in number of occupied hectads between the two time periods indicating distributional change. Symbols next to the species name highlight cases where the current time period differs from the standard definition of 2000–16 (+ indicates 2005–16 and * indicates 2010–16). Where species are non-native (NN), are reintroduced, or are naturalised (nat), this is shown in parentheses next to the common name. Naturalised species are those that are present in the UK owing to human activities, but have been present since at least Roman times.

Common name	Records 1960–92	Records 2000–16	Hectads 1960–92	Hectads 2000–16	Hectad change
Hedgehog	17,618	134,753	2,191	2,507	316
European mole	21,807	65,493	2,259	2,325	66
Common shrew	7,449	8,514	1,630	1,445	**−185**
Pygmy shrew	3,374	4,218	1,195	1,008	**−187**
Water shrew	1,946	3,318	834	972	138
Lesser white-toothed shrew (nat)	38	64	4	4	0
European rabbit (nat)	21,251	106,569	2,491	2,540	49
Brown hare (nat)	13,948	49,378	1,847	1,967	120
Mountain hare	2,186	5,945	475	607	132
Red squirrel (*)	9,329	27,565	1,193	806	**−387**
Grey squirrel (*, NN)	14,833	42,575	1,511	1,772	261
Eurasian beaver (reintroduced)	0	164	0	53	53
Hazel dormouse	2,717	30,898	453	572	119
Edible dormouse (NN)	128	154	13	23	10
Bank vole	4,736	9,156	1,258	1,284	26
Field vole	6,036	12,340	1,482	1,624	142
Common vole (nat)	46	45	17	9	**−8**
Water vole (+)	9,879	32,764	1,326	1,224	**−102**
Harvest mouse	2,566	2,926	788	649	**−139**
Wood mouse	6,807	20,296	1,627	1,641	14
Yellow-necked mouse	693	3,467	271	371	100
House mouse (nat)	3,242	2,542	1,074	686	**−388**
Brown rat (NN)	6,933	16,824	1,621	1,561	**−60**
Black rat (nat)	159	29	82	16	**−66**
Wildcat	602	371	351	111	**−240**
Red fox	13,686	91,616	2,035	2,452	417
Badger	19,825	80,110	2,048	2,292	244
Otter	24,760	82,866	2,051	2,394	343

Common name	Records 1960–92	Records 2000–16	Hectads 1960–92	Hectads 2000–16	Hectad change
Pine marten	767	3,268	329	559	230
Stoat	7,645	13,675	1,791	1,771	**−20**
Weasel	7,565	8,735	1,664	1,527	**−137**
Polecat	1,915	4,210	332	775	443
American mink (NN)	5,311	12,213	1,402	1,617	215
Wild boar (reintroduced)	5	468	3	67	64
Red deer	4,827	41,262	970	1,222	252
Sika deer (NN)	850	8,488	266	411	145
Fallow deer (nat)	2,701	19,507	636	1,028	392
Roe deer	8,234	87,385	1,470	2,091	621
Chinese water deer (NN)	400	3,251	71	120	49
Reeves' muntjac deer (NN)	2,998	20,622	680	966	286
Greater horseshoe bat	4,121	5,451	192	293	101
Lesser horseshoe bat	8,196	17,856	334	460	126
Alcathoe bat	0	35	0	19	19
Whiskered bat	1,357	3,018	446	725	279
Brandt's bat	337	798	112	241	129
Bechstein's bat	174	1,468	33	134	101
Daubenton's bat	3,506	24,521	655	1,479	824
Greater mouse-eared bat	95	78	5	1	**−4**
Natterer's bat	3,389	13,759	679	1,190	511
Serotine bat	1,420	9,304	300	708	408
Leisler's bat	463	1,419	124	453	329
Noctule bat	1,834	16,199	531	1,362	831
Common pipistrelle bat	3,037	63,549	527	1,999	1,472
Soprano pipistrelle bat	0	37,920	0	1,848	1,848
Nathusius' pipistrelle bat	12	1,363	11	432	421
Barbastelle bat	144	2,815	82	436	354
Brown long-eared bat	8,853	29,375	1,378	1,768	390
Grey long-eared bat	72	246	33	45	12
Grey seal	1,920	11,058	643	1,007	364
Harbour seal	995	6,245	450	725	275

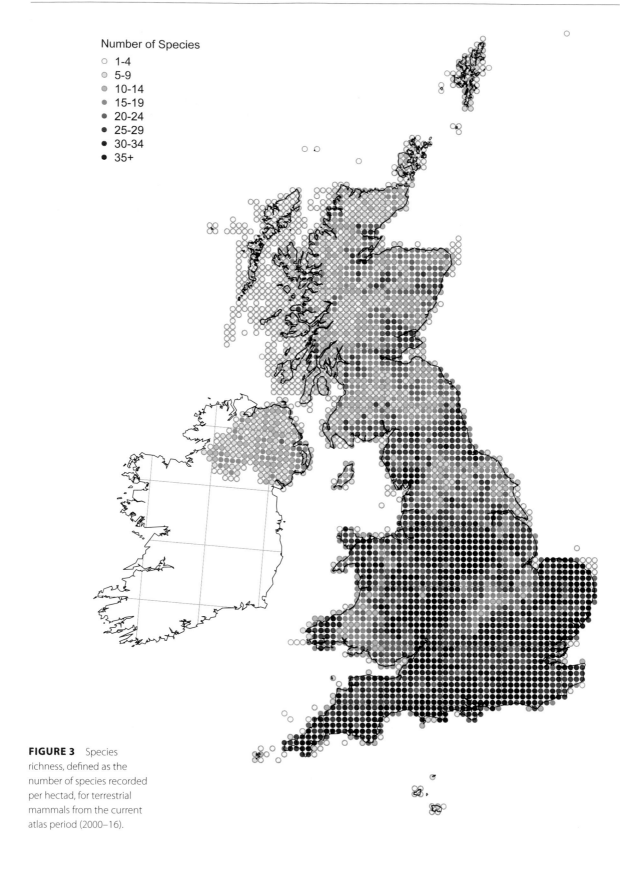

Number of Species

○ 1-4
○ 5-9
○ 10-14
● 15-19
● 20-24
● 25-29
● 30-34
● 35+

FIGURE 3 Species richness, defined as the number of species recorded per hectad, for terrestrial mammals from the current atlas period (2000–16).

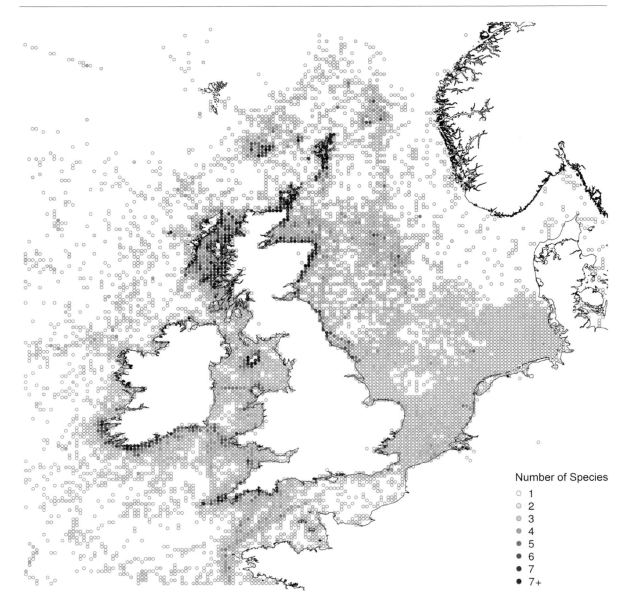

FIGURE 4 Species richness of cetaceans, calculated as the number of species recorded per hectad from the current period (2000–2016).

Number of Species

○ 1
◔ 2
◑ 3
◕ 4
● 5
● 6
● 7
● 7+

deer and bat species have expanded their distribution. However, it can be difficult to determine whether the observed change of distribution reflects changes in observer effort (all), increases in recording technology (bats) and better species identification (bats), or an actual biological change in distribution (mustelids).

The terrestrial species richness across the UK, defined here as the number of different species within each hectad, ranges from one to over 35 in the current time period (Figure 3). There is a distinct pattern, with the hectads in the south of Britain having a higher species richness than those in northern England, north and west Scotland, and Northern Ireland.

For cetaceans, the number of species recorded per cell is greatest along Atlantic coasts (Figure 4): the Hebrides, southern Ireland, south-west England, and in the Irish Sea around Pembrokeshire, the Llŷn Peninsula, Anglesey and the Isle of Man. In the North Sea, the greatest number of species occurs in the north-west

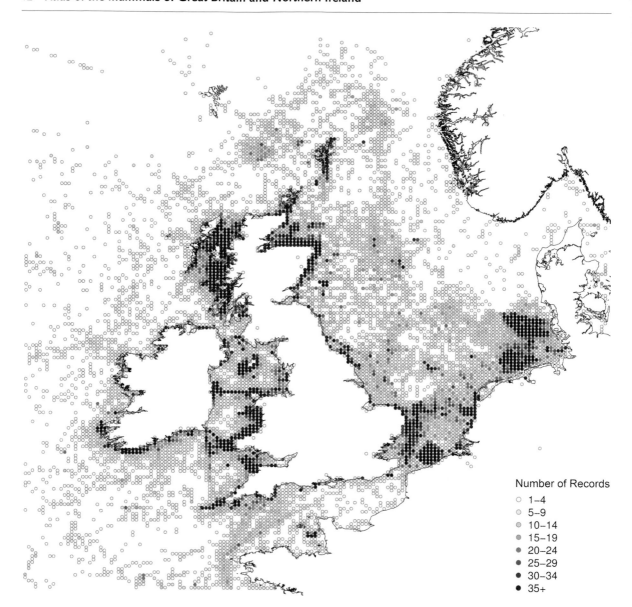

Number of Records
○ 1–4
◔ 5–9
◑ 10–14
◕ 15–19
● 20–24
● 25–29
● 30–34
● 35+

FIGURE 5 The number of cetacean records per hectad from the current atlas period from across the North Sea and within the shelf seas surrounding Britain and Ireland.

of the region, around north and east Scotland. These are also the areas where the greatest number of records occur (Figure 5), influenced to some extent by variation in observer effort. However, sightings rates and species diversity from effort-corrected surveys show a similar pattern.

BIBLIOGRAPHY

ARNOLD, H.R. 1993. *Atlas of Mammals in Britain*. ITE Research Publication No. 6. London: Joint Nature Conservation Committee and Institute of Terrestrial Ecology, HMSO.

BIVAND, R. and LEWIN-KOH, N. 2015. Maptools: Tools for Reading and Handling Spatial Objects. R package version 0.8–36.

HIJMANS, R.J. 2013. Raster: Geographic data analysis and modelling. R package version 2.1–66.

R CORE TEAM. 2017. R: A language and environment for statistical computing. R Foundation for Statistical Computing, Vienna, Austria. Accessed at https://www.R-project.org/ (30 July 2019).

SPECIES
ACCOUNTS

Hedgehog
Erinaceus europaeus (LINNAEUS, 1758)

MARK BALDWIN

DISTRIBUTION

The hedgehog remains widespread in suitable habitats across the UK, but there is good evidence of a serious and continuing decline. Many possible factors can be potentially associated with this change in abundance, including agricultural intensification (particularly the use of biocides and a reduction in the availability of prey, nest sites and refuges), habitat loss and degradation, landscape fragmentation, and mortality from road traffic and predation. This species is generally absent from wetlands, moors, uplands, coniferous forests and city centres.

Hedgehogs have been introduced to many archipelagos and islands: Shetland, Orkney, the Western Isles (Lewis, Harris, Benbecula, North and South Uist), Skye, Soay, Canna, Coll, Mull, Ulva, Luing, Islay, Arran, Bute, the Isle of Man, Anglesey, the Isle of Wight, St Mary's (Isles of Scilly), Alderney, Guernsey, Jersey and Sark.

They were probably introduced to mainland Ireland and Beginish Island (Kerry).

ECOLOGY

The hedgehog is a nocturnal insectivore found in a wide range of rural and urban environments. It eats a variety of macroinvertebrates including worms, beetles, insect larvae and gastropod molluscs, but it may also take carrion and small vertebrates. The species spends the day in a nest in undergrowth which is typically constructed of broad leaves and/or grasses. The hedgehog hibernates in winter.

IDENTIFICATION

The hedgehog is the UK's only spiny mammal. Sharp spines completely replace hair on the back and crown of the head. It 'rolls up' in defence leaving only spiny skin exposed. The pelage is typically mid-brown. It has a short tail, small eyes and a tapering snout.

BIBLIOGRAPHY

HOF, A.R. 2009. A study of the current status of the hedgehog (*Erinaceus europaeus*), and its decline in Great Britain since 1960. Unpublished PhD thesis, Royal Holloway University of London.

MORRIS, P.A. & REEVE, N.J. 2008. Hedgehog *Erinaceus europaeus*. In S. Harris & D.W. Yalden (eds) *Mammals of the British Isles*, 4th edn. Southampton: The Mammal Society. pp. 241–249.

PTES & BHPS. 2015. *The State of Britain's Hedgehogs 2015*. London and Ludlow: People's Trust for Endangered Species and British Hedgehog Preservation Society.

AUTHOR   Nigel Reeve

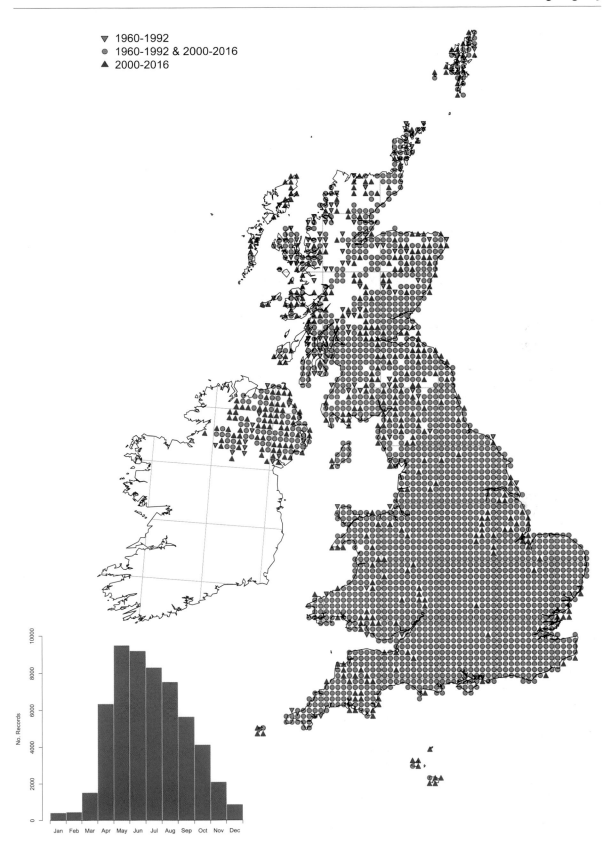

European mole

Talpa europaea (LINNAEUS, 1758)

ANDREW FUSEK PETERS

DISTRIBUTION

The mole is found throughout Great Britain, including on several islands where it was probably introduced. However, it is absent from Ireland. It is found in most habitats where the soil is deep enough to allow tunnelling, but is uncommon in coniferous forests, on moorlands and in sand dunes, probably because its prey is scarce here. There has been no change in distribution over time, even though it is regularly trapped. Most records are from molehills, with some live sightings and dead animals, including those killed by cats.

ECOLOGY

The mole spends almost all of its life underground in a system of permanent and semi-permanent tunnels. Surface tunnels are usually short-lived and are employed in very shallow soils where prey is concentrated just below the surface. A system of permanent deep burrows, which form a complex network hundreds of metres long at varying soil depths, is more common. Permanent tunnels are used repeatedly for feeding over long periods of time, sometimes by several generations of moles. Molehills are pushed up when redigging trampled tunnels or when establishing new territories. Moles are often seen in spring and autumn, associated with the onset of breeding and the dispersal of juveniles.

IDENTIFICATION

The mole has short velvety fur. Its fur is usually black, silver, grey, cream or white, but shades of brown have also been recorded and can dominate in specific populations. It has spade-like forelimbs with large claws that face towards the rear of the animal. It has a pink fleshy snout and tiny eyes.

BIBLIOGRAPHY

ATKINSON, R. 2013. *Moles*, 1st edn. London: Whittet Books. STONE, D. 1986. *Moles*. Oswestry: Anthony Nelson.

AUTHOR Derek Crawley

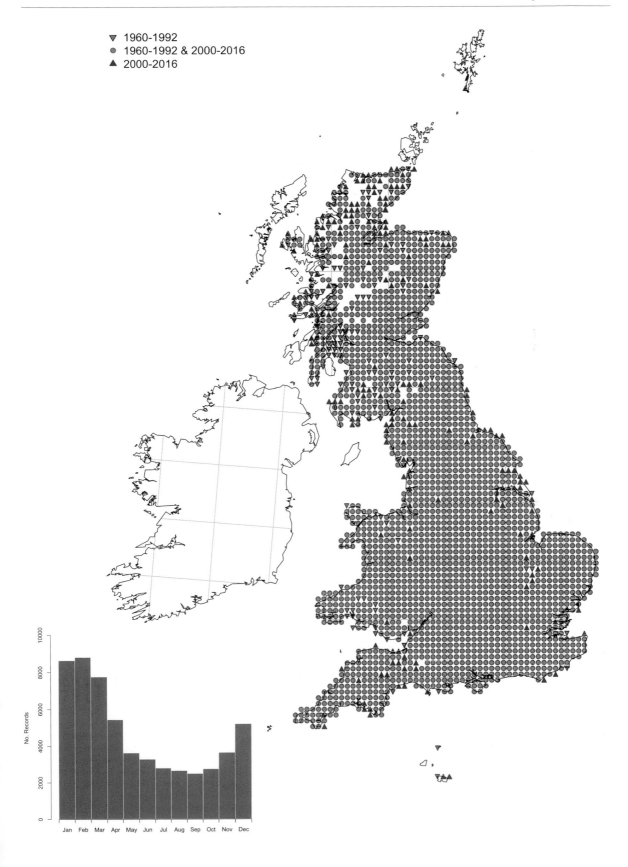

▽ 1960-1992
● 1960-1992 & 2000-2016
▲ 2000-2016

Common shrew
Sorex araneus (LINNAEUS, 1758)

MALCOLM WELCH

aggressive towards each other. The species is active almost continuously throughout the day and night.

IDENTIFICATION

The common shrew has a long pointed nose, small eyes, small ears and red-tipped teeth. It has tricoloured fur: dark brown on the back, pale brown on the sides and whitish underneath. It can be distinguished from the water shrew *Neomys fodiens* by its brown (not black) upper surface, and smaller size. The common shrew differs from the smaller pygmy shrew *Sorex minutus* by having distinctively sharp contrast between the coat colours on the back and flank, as well as by having a relatively shorter, thinner tail.

BIBLIOGRAPHY

CHURCHFIELD, S. & SEARLE, J. 2008. Common shrew *Sorex araneus*. In S. Harris & D.W. Yalden (eds) *Mammals of the British Isles*, 4th edn. Southampton: The Mammal Society. pp. 257–265.

AUTHOR   Anna Champneys

DISTRIBUTION

The common shrew is widespread throughout much of Great Britain. However, it is absent from Ireland, as well as from Shetland, Orkney, the Outer Hebrides and the Isles of Scilly. Records come mainly from cat kills, animals live-trapped during small mammal surveys, and remains found in barn owl pellets.

ECOLOGY

The common shrew is found in most terrestrial habitats where there is short vegetation cover such as thick grass, bushy scrub, hedgerows, bracken, deciduous woodland and roadside verges. It is an abundant species, with highest population densities being recorded in grassland habitats and bog. Its diet consists mainly of terrestrial invertebrates such as earthworms, slugs, snails, beetles, spiders and woodlice. Essentially solitary except during the breeding season, individuals are extremely

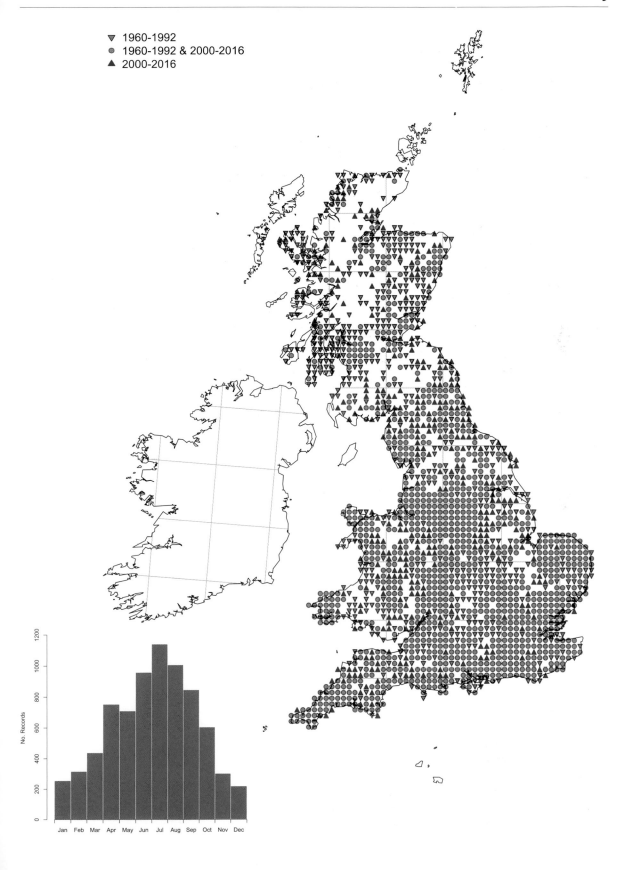

Legend:
- ▽ 1960-1992
- ● 1960-1992 & 2000-2016
- ▲ 2000-2016

Pygmy shrew
Sorex minutus (LINNAEUS, 1766)

BECKY HAYWOOD

DISTRIBUTION

The pygmy shrew is widespread throughout Great Britain and Ireland. However, it is absent from Shetland, Lewis, the Isles of Scilly and the Channel Islands. Records are scarce and come mainly from cat kills, animals live-trapped during small mammal surveys, and remains found in barn owl pellets.

ECOLOGY

The pygmy shrew is found in most terrestrial habitats where there is plenty of ground cover. Population densities are typically higher in grassland than woodland with the exception of Ireland, where its competitor – the common shrew – is absent. The pygmy shrew is less abundant than the common shrew, and has a similar diet, although with a tendency for smaller-sized prey. Spiders, harvestmen, woodlice and beetles are its most common prey. The pygmy shrew, like the common shrew, is essentially solitary except during the breeding season and individuals are extremely aggressive towards each other. The species is active almost continuously throughout the day and night.

IDENTIFICATION

The pygmy shrew is similar in appearance to the common shrew with a long pointed nose, small eyes, small ears and red-tipped teeth. It is, however, much smaller and possesses a relatively longer, thicker and more hairy tail. Its coat is also bicoloured, lacking the distinctive sharp contrast between coat colour on the back and flank seen in the common shrew.

BIBLIOGRAPHY

CHURCHFIELD, S. & SEARLE, J. 2008. Pygmy shrew *Sorex minutus*. In S. Harris & D.W. Yalden (eds) *Mammals of the British Isles*, 4th edn. Southampton: The Mammal Society. pp. 267–271.
MACDONALD, D.W., MACE, G.M. & RUSHTON, S. 1998. *Proposals for Future Monitoring of British Mammals*. London: Department of the Environment, Transport and the Regions with Joint Nature Conservation Committee, Peterborough.

AUTHOR   Anna Champneys

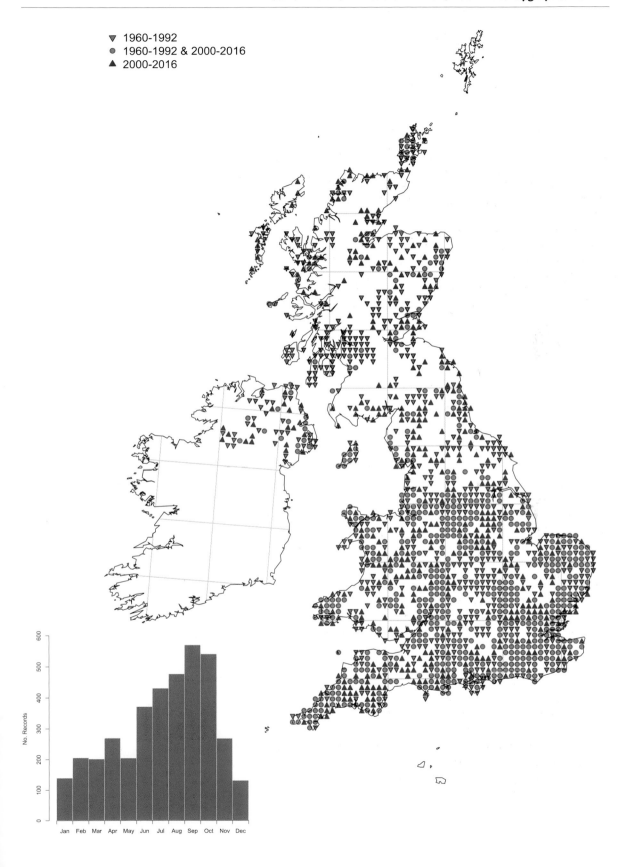

Water shrew
Neomys fodiens (PENNANT, 1771)

KATIE NETHERCOAT

DISTRIBUTION

Within Great Britain, the water shrew has a wide but patchy distribution. It is present on many islands including Skye, Mull, Anglesey and the Isle of Wight. However, it is more localised in Scotland and is absent from Ireland. It typically occurs at much lower population densities than its terrestrial counterparts, the common and pygmy shrews.

The first national survey in 2004–05 identified water shrew presence in riparian habitats by the analysis of faecal samples collected in baited tubes. Of the 2,159 sites surveyed, 17% showed signs of water shrew occurrence. Although the presence of water shrews largely reflected the distribution of surveyors (the majority coming from England (82%), with 10% from Scotland and 8% from Wales), there was a notable concentration of positive records in central and eastern England.

Records come mainly from cat kills, animals live-trapped during small mammal surveys, bait tube surveys, and remains found in barn owl pellets.

ECOLOGY

The water shrew occupies a wide variety of riparian habitats including ponds, canals, reed beds and marshes as well as clear, fast-flowing streams and watercress beds – habitats with which it has been traditionally associated. However, it is also found regularly, if infrequently, away from water in agricultural habitats. Its diet consists mainly of terrestrial and aquatic invertebrates.

The water shrew, like its terrestrial shrew counterparts, is solitary and territorial with an elusive nature. It exists in small localised populations and is active throughout the day and night.

IDENTIFICATION

The water shrew is the largest of the six shrews inhabiting Britain; it can be distinguished from the terrestrial shrews by its short, dense, velvety black fur on the upper surface of the body and greyish-white underside. Most water shrews have tufts of white fur on the ears and around the eyes. There are distinctive stiff hairs on the edges of the feet and along the underside of the tail that form a keel. The water shrew has a long pointed nose, small eyes and small ears.

BIBLIOGRAPHY

CARTER, P. & CHURCHFIELD, S. 2006. *Distribution and Habitat Occurrence of Water Shrews in Great Britain.* Environment Agency Science Report SC010073/SR. Bristol: Environment Agency. Mammal Society Research Report No. 7. London: The Mammal Society.

CHURCHFIELD, S. 2008. Water shrew *Neomys fodiens*. In S. Harris & D.W. Yalden (eds) *Mammals of the British Isles*, 4th edn. Southampton: The Mammal Society. pp. 271–275.

AUTHOR   Anna Champneys

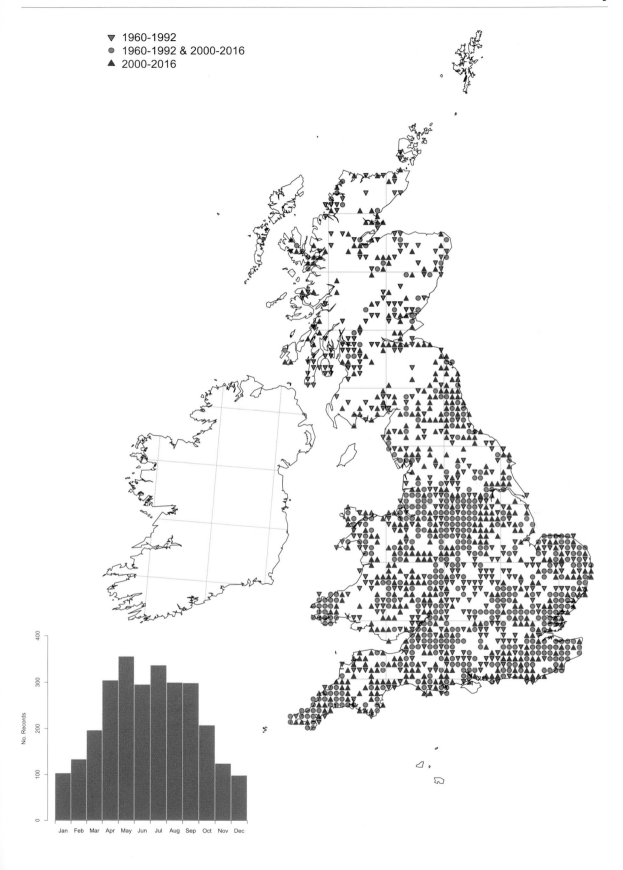

Lesser white-toothed shrew
Crocidura suaveolens (PALLAS, 1811)

JULIE LOVE

DISTRIBUTION

In the UK, the lesser white-toothed shrew is found only on the Isles of Scilly – where it is most abundant on St Mary's and Tresco – and on the Channel Islands. Records come from animals live-trapped during small mammal surveys, faecal remains in bait tubes, cat kills, incidental sightings and individuals found under corrugated iron sheets.

ECOLOGY

The lesser white-toothed shrew is found in most terrestrial habitats where there is adequate cover such as bracken, hedgerows and woodland. It can also be found in coastal habitats among boulders and vegetation on the shores. The most common prey of this species includes beetles, flies, insect larvae, centipedes, earthworms and gastropods. However, in coastal habitats, it feeds predominantly on crustaceans. The lesser white-toothed shrew is generally solitary but it has overlapping home ranges, so it is not as territorial as the other shrews.

The species is active almost continuously throughout the day and night.

IDENTIFICATION

The lesser white-toothed shrew has a long pointed nose, small eyes, short-haired prominent ears and wholly white teeth. Its coat is bicoloured with greyish or reddish brown fur on the back and paler fur on the underside. The tail of this species is covered with short hair interspersed with long, white hairs. It is the only shrew species on the Isles of Scilly. On Jersey it can be distinguished from the Millet's shrew *Sorex coronantus* by the lack of red-tipped teeth and its long, scattered tail hairs.

BIBLIOGRAPHY

CHURCHFIELD, S. & TEMPLE, R.K. 2008. Lesser white-toothed shrew *Crocidura suaveolens*. In S. Harris & D.W. Yalden (eds) *Mammals of the British Isles*, 4th edn. Southampton: The Mammal Society. pp. 276–280.

CORNWALL MAMMAL GROUP. 2015. Lesser white-toothed shrew. Accessed at https://www.cornwallmammalgroup.org/lesser-white-toothed-shrew (30 July 2019).

AUTHOR   Anna Champneys

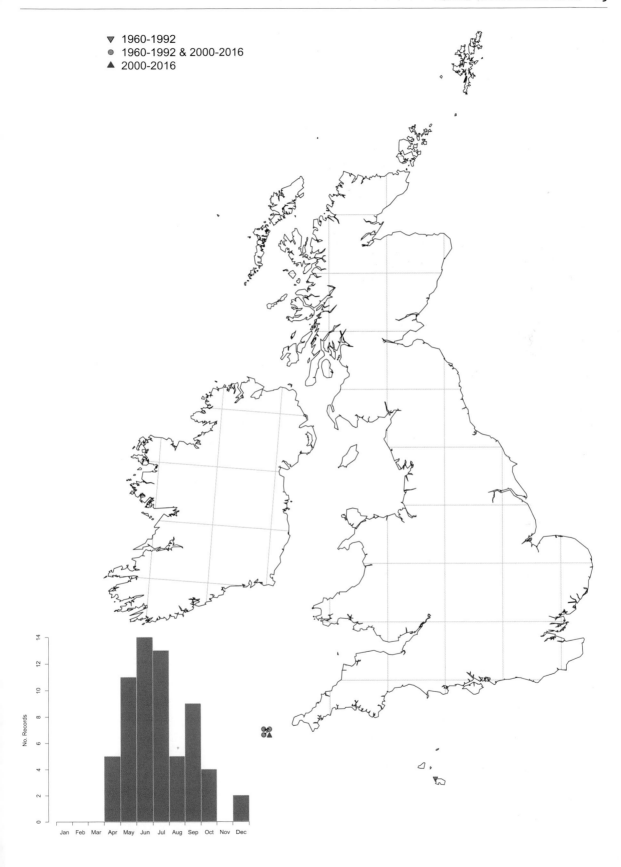

European rabbit
Oryctolagus cuniculus (LINNAEUS, 1758)

PADRAIG KAVANAGH

DISTRIBUTION

The rabbit is a naturalised species with a wide distribution across the UK. It is recorded in more hectads than any other species in both the current and historical atlas periods. However, there are currently concerns about its population status in parts of its range, particularly in Scotland, based on data collected by the British Trust for Ornithology's Breeding Bird Survey. Mass mortalities have been caused by myxomatosis and rabbit viral haemorrhagic disease, but there remains uncertainty in the scale of the decline owing to the large natural variability in rabbit population sizes and the lack of systematic monitoring for this species.

ECOLOGY

The rabbit has a herbivorous diet that can include a range of plant species. Where available, it prefers short grass swards where mammalian predators are visible. Tree bark may be consumed during winter months, and fallen leaves during summer droughts. The rabbit lives in warren-based territorial social groups of 1–3 males and 1–9 females in areas of burrowable substrate. The dispersal of juveniles is male-biased, with female offspring remaining in their natal social groups.

The rabbit may be active above ground throughout the day, although activity peaks during crepuscular periods in areas of human disturbance. The species has become recognised as a keystone ecosystem engineer providing environmental conditions required by a range of fauna and flora in UK habitats such as the East Anglian Breckland and southern chalk downland. It is also prey for a number of avian and mammalian predators.

IDENTIFICATION

The species is easily distinguished from the brown hare *Lepus europaeus* by its shorter limbs and shorter ears, which lack the black tip of the latter. The brown pelage on the top of the body can be replaced by black coat in melanistic individuals in some regions. The adult body weight of a UK wild rabbit can reach 2 kg.

BIBLIOGRAPHY

HARRIS, S.J., MASSIMINO, D., GILLINGS, S., EATON, M.A., NOBLE, D.G., BALMER, D.E., PROCTER, D., PEARCE-HIGGINS, J.W. & WOODCOCK, P. 2018. *The Breeding Bird Survey 2017*. The British Trust for Ornithology Research Report 706. Thetford: British Trust for Ornithology.

LEES, A.C. & BELL, D.J. 2008. A conservation paradox for the 21st century: the European wild rabbit *Oryctolagus cuniculus*, an invasive alien and an endangered native species. *Mammal Review* 38 (4): 304–320.

WEBB, N.J., IBRAHIM, K., BELL, D.J., HEWITT, G. 1995. Natal dispersal and genetic structure in a population of the European wild rabbit *Oryctolagus cuniculus*. *Molecular Ecology* 4 (2): 239–247.

AUTHOR   Diana Bell

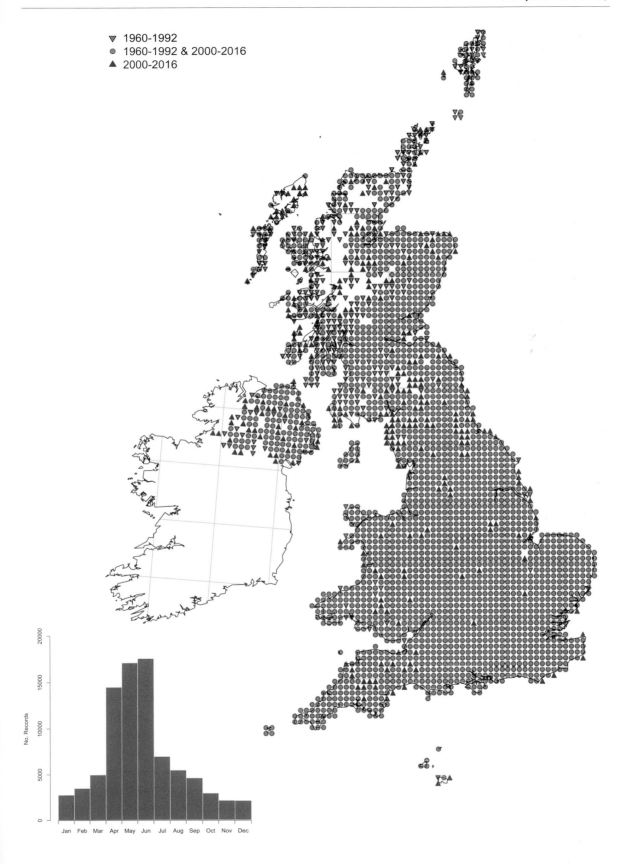

Brown hare
Lepus europaeus (PALLAS, 1778)

IAN HULL

DISTRIBUTION

There were no records of the brown hare in Britain before the Roman invasion in the first century AD, and it was introduced to Ireland for sport in the mid-late nineteenth century. Since its introduction, it has replaced the indigenous mountain hare *Lepus timidus* across most of Britain, and competes with it in Ireland. However, it is unable to eat coarse food and this has led to it staying below the heather line. Brown hare population sizes are thought to have dropped by 90% since the second half of the nineteenth century.

ECOLOGY

The brown hare breeds throughout most of the year. The female hare, or jill, can have up to four litters a year, although most do not breed until their second year. Boxing, the jill rejecting the male, is most often noted in spring when the crops are low and most females come into season. The brown hare evolved on steppes and plains and spread as humans cut down forests. It prefers old meadow land with a mix of wild plants; leverets cannot eat farm crops when they get coarse.

Hares produce caecotrophs from the anus which they eat, thus 'double-digesting' their food.

IDENTIFICATION

The brown hare can be distinguished from the rabbit by its larger size and its longer ears with distinctive black tips. An adult brown hare weighs 3.5 kg. It is a tall animal with long legs that lopes or runs with a well-developed stride. At full gallop, a brown hare can reach speeds of 55–70 km/h (35–45 mph). When running, the white tail is usually held downwards showing a black upper surface. The coat is brown, ranging from russet to dark brown, with white on the belly, and the iris of the eye is orange. However, hair and eye colour mutations are found.

BIBLIOGRAPHY

RUSS, J. 2015. *The Hare Book*. Llangennech: Hare Preservation Trust and Graffeg Publishing.
TAPPER, S. & YALDEN, D. 1987. *The Brown Hare*. Southampton: The Mammal Society.

AUTHOR   Jo Sharplin

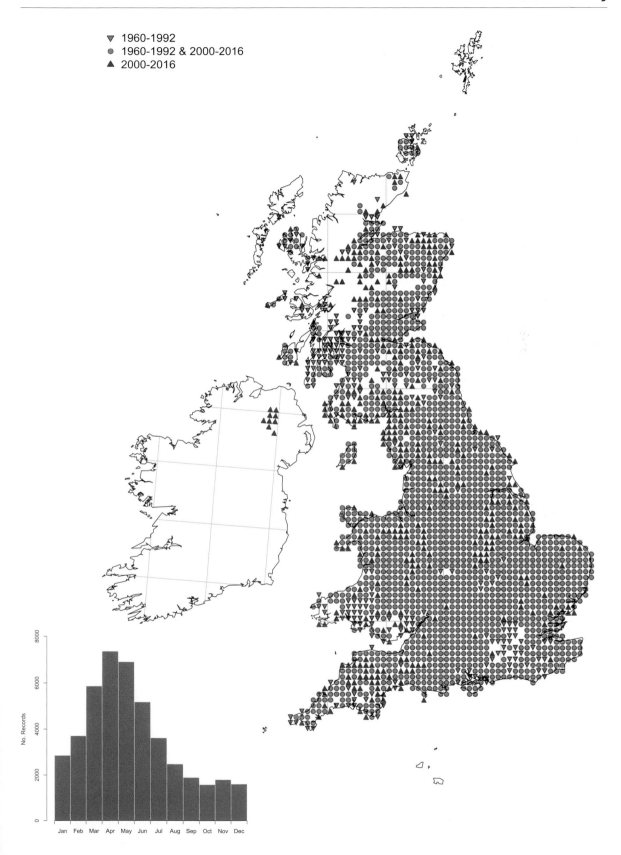

Mountain hare
Lepus timidus (LINNAEUS, 1758)

BILLY STOCKWELL

DISTRIBUTION

The mountain hare is the indigenous hare that has been in Britain since the last Ice Age. It is native to the Highlands of Scotland, but was introduced by the Victorians to the Peak District in Derbyshire. It is also found on some Scottish islands including Hoy (Orkney), Shetland, Mull and Skye. The subspecies *L. timidus hibernicus* is indigenous to Ireland, where it is known as the Irish hare.

It is associated with heather moorlands and occupies higher ground than the brown hare: some 350 m up in the Lammermuir Hills in the Scottish Borders and 750 m to over 1,000 m in the Cairngorms. Population densities are known to vary at least tenfold, reaching a peak roughly every ten years; reasons for these variations are uncertain. There is concern that high culling rates in some areas may adversely affect the population status of the species, but the high variability in population densities, particularly in habitats where they are most abundant such as grouse-moors, makes understanding population trends challenging.

ECOLOGY

As well as heather, the mountain hare will browse gorse, and dwarf shrubs or trees such as birch, rowan and juniper. In summer, it prefers to eat grasses when available. In its upland and moorland habitat, it makes forms – depressions in snow, soil, or in the lea of a heather hummock – which provide shelter. It rests in these by day and at night, and in periods of snow cover it often gathers in large groups on leeward slopes to shelter or to feed. Its 'runs' typically pass up slopes as the mountain hare's short front legs and powerful long back legs are ideal for running uphill at speeds of up to 60 km/h (37 mph). Mating begins at the end of January, and gestation lasts about 50 days. Most leverets are born between March and August inclusive, fully furred and with eyes open. There can be several litters a year, usually between one and three young. On average, the lifespan of the mountain hare is three to four years.

IDENTIFICATION

The mountain hare is smaller and more compact than the brown hare; its total body length can range from 43 cm to 61 cm, and it weighs 2.5–3.5 kg. Its black-tipped ears, at 6–8 cm, are also shorter than those of the brown hare. In spring it loses its thick white coat, changing to a brownish one. The brown coat begins to shed in late October, and by December it again has thick white fur. The tail remains white, and the ear tips black, throughout the seasons. The moult is dependent upon temperature and appears to be initiated by shortening day lengths.

BIBLIOGRAPHY

HESFORD, N., FLETCHER, K., HOWARTH, D., SMITH, A.A., AEBISCHER, N.J. & BAINES, D. 2019. Spatial and temporal variation in mountain hare *(Lepus timidus)* abundance in relation to red grouse *(Lagopus lagopus scotica)* management in Scotland. *European Journal of Wildlife Research* 65 (3): 33.

RUSS, J. 2015. *The Hare Book*. Llangennech: Hare Preservation Trust and Graffeg Publishing.

WATSON, A. & WILSON, J.D. 2018. Seven decades of mountain hare counts show severe declines where high-yield recreational game bird hunting is practised. *Journal of Applied Ecology* 55 (6): 2663–2672.

AUTHOR   Marion O'Neil and Jo Sharplin

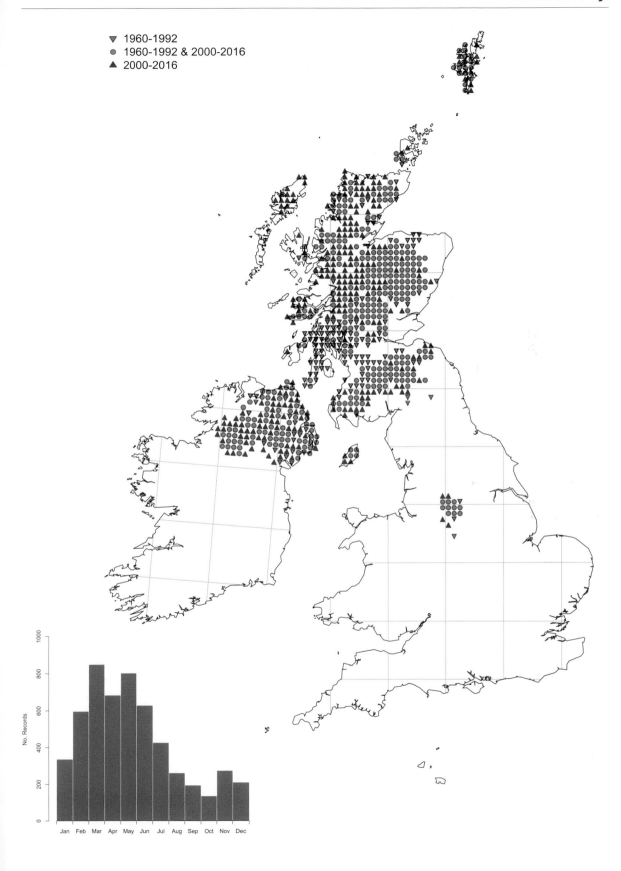

▽ 1960-1992
● 1960-1992 & 2000-2016
▲ 2000-2016

Red squirrel
Sciurus vulgaris (LINNAEUS, 1758)

PADRAIG KAVANAGH

DISTRIBUTION

The red squirrel is native to Britain and Ireland. Historically eliminated from some areas by culling, restocking in the nineteenth century included animals taken from populations elsewhere in Europe. It has recently been introduced to Tresco, Mersea and Caldey, and a series of conservation-driven introductions have occurred in the north and north-west of Scotland, beyond its historical range.

Over recent decades, widespread regional extinctions of the red squirrel have occurred as a result of resource competition with, and epizootic infection spread by, the grey squirrel. While absent from Scotland's central belt, significant mainland populations remain in northern and border counties. The species also remains widespread in Cumbria and Northumberland, while small remnant populations persist in Durham, Yorkshire and Lancashire. Recent records in the southern part of England are likely to have originated from escapes. In Wales, isolated populations are found in northern Gwynedd, Clocaenog forest (Denbighshire) and conifer plantations surrounding Llyn Brianne (Tywi, Crychan forests) in mid-Wales. Notable island populations include Anglesey, Arran, Brownsea, Furzey and the Isle of Wight. In Ireland, the natural re-establishment of pine martens *Martes martes* is correlated with grey squirrel decline and red squirrel recovery. In the absence of this predator, Irish red squirrel populations continue to decline as the grey squirrel increases it range.

Recent survey and data review suggest that grey squirrel control may have stabilised, and in some localities begun to reverse, regional red squirrel declines. However, the pattern is complex, and most populations remain at risk from sympatric or adjacent grey squirrel populations.

ECOLOGY

The red squirrel is an arboreal diurnal species typical of mature coniferous and broadleaved woodlands. Its diet is largely tree seeds, flowers and shoots, with a little animal matter and fungi.

IDENTIFICATION

The red squirrel has a long tail and typically prominent ear tufts. It has white chest hair, but the rest of the coat can vary from russet red to brown, black and occasionally grey. It can be confused with the larger and stockier grey squirrel, which seldom has ear tufts and has a characteristic outer band of silver on the tail hair. The red squirrel is usually smaller than the grey squirrel, with a body length of 18–23 cm.

BIBLIOGRAPHY

GURNELL, J., LURZ, P. & BERTOLDI, W. 2014. The changing patterns in the distribution of red and grey squirrels in the North of England and Scotland between 1991 and 2010 based on volunteer surveys. *Hystrix, the Italian Journal of Mammalogy* 25 (2): 83–89.
LAWTON, C., FLAHERTY, M., GOLDSTEIN, E.A., SHEEHY, E. & CAREY, M. 2012. *Irish squirrel survey 2012*. Irish Wildlife Manuals, No. 89. Dublin: National Parks and Wildlife Service, Department of Arts, Heritage and the Gaeltacht, Ireland.
SHUTTLEWORTH, C.M., LURZ, P. & HAYWARD, M.W. 2015. *Red Squirrels: Ecology, Conservation and Management in Europe*. Kenilworth: European Squirrel Initiative.

AUTHOR   Craig Shuttleworth

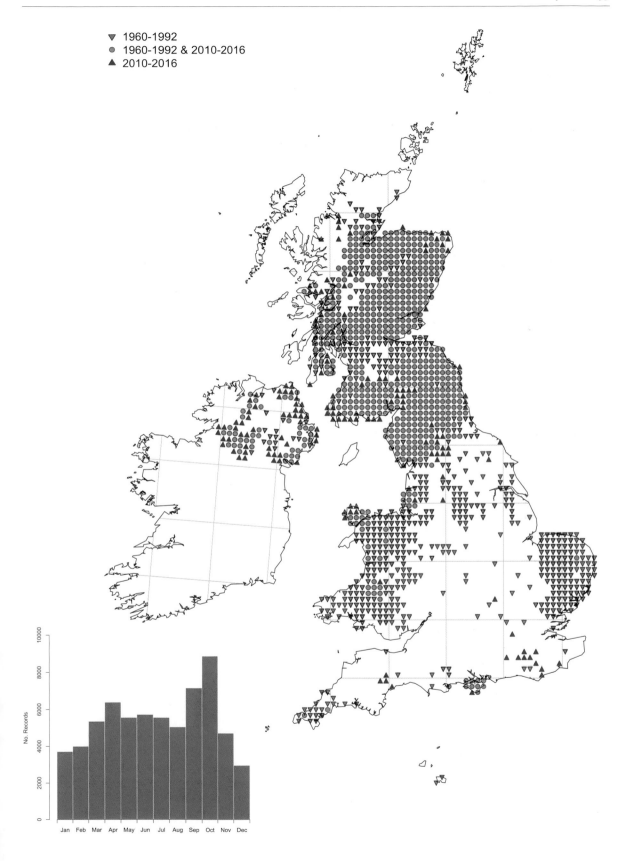

- ▽ 1960-1992
- ⬤ 1960-1992 & 2010-2016
- ▲ 2010-2016

Grey squirrel
Sciurus carolinensis (GMELIN, 1788)

ROBIN MORRISON

DISTRIBUTION

The grey squirrel was introduced to the UK in 1876, and to Ireland in 1911, from North America. It is widely distributed and common throughout England, Wales, much of Ireland, southern Scotland and the central belt. Natural range expansion continues, with the exception of the Irish midlands, where population declines are correlated with the establishment of high-density pine marten populations. No populations exist on offshore islands since the species was eradicated from Anglesey in 2013. Individuals have, however, occasionally been found on the island of Skye (accidental transportation), the Isle of Wight (accidental or deliberate transportation) and Anglesey (limited natural dispersal from mainland).

Systematic 4 km² presence/absence surveys are carried out in Scotland and northern England as part of red squirrel conservation measures. These regional assessments are supplemented with records of culled animals and sighting reports. Elsewhere, because the grey squirrel lives at high density and is a highly visible species, distribution is not monitored through systematic study. Populations are known to be affected by both weather patterns and annual availability of tree seed crops; supplemental feeding (at bird tables) may thus buffer fluctuation in local population abundance.

Deliberate or accidental transportation and release could accelerate range expansion. It is important that potential sightings of grey squirrels beyond the current established distribution are reported.

ECOLOGY

The grey squirrel is a diurnal species, less arboreal than the red squirrel and able to occupy more fragmented woodland habitats. Its diet is principally tree seeds, flowers and shoots, with some animal matter and fungi; it also strips bark and because of this is considered an economic pest.

IDENTIFICATION

The grey squirrel is larger and stockier (body length 23–27 cm) than the native red squirrel and seldom has ear tufts. It also typically has a silver-coloured fringe to the tail hair and, although most individuals are grey in colour, many have reddish pelage on head and flanks, with some animals being very red. Melanistic (black) individuals can be found in south-east England.

BIBLIOGRAPHY

GURNELL, J., LURZ, P. & BERTOLDI, W. 2014. The changing patterns in the distribution of red and grey squirrels in the North of England and Scotland between 1991 and 2010 based on volunteer surveys. *Hystrix, the Italian Journal of Mammalogy* 25 (2): 83–89.

LAWTON, C., FLAHERTY, M., GOLDSTEIN, E.A., SHEEHY, E. & CAREY, M. 2012. *Irish squirrel survey 2012.* Irish Wildlife Manuals, No. 89. Dublin: National Parks and Wildlife Service, Department of Arts, Heritage and the Gaeltacht, Ireland.

O'HARE, S. 2017. *Results of the squirrel monitoring programme, Spring 2017.* Carlisle: Red Squirrels Northern England.

AUTHOR   Craig Shuttleworth

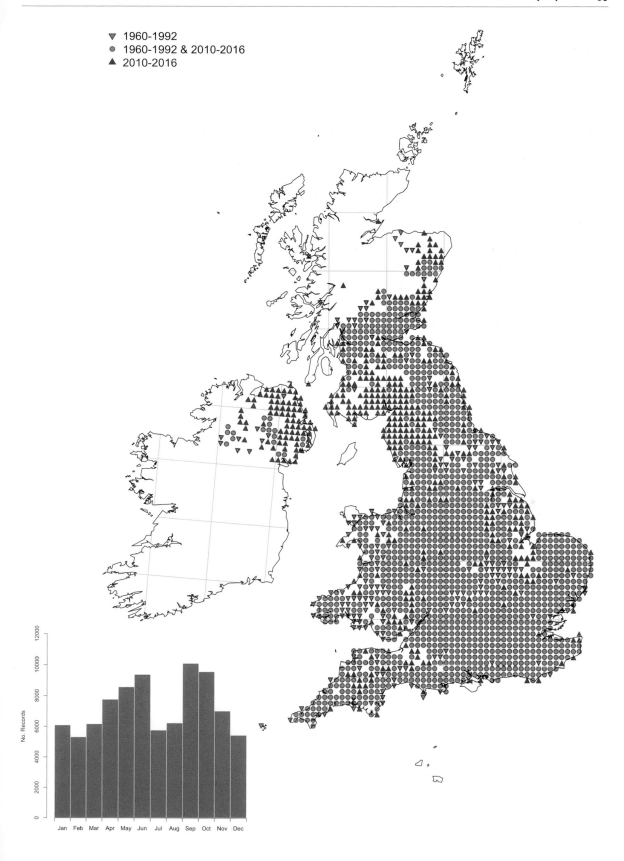

Eurasian beaver
Castor fiber (LINNAEUS, 1758)

TONY HOUSE

DISTRIBUTION

The Eurasian beaver is recovering from near global extinction and is now present throughout its former native range thanks to conservation efforts, including legal protection and translocations to over 25 European countries. Britain was one of the last EU states to consider restoring this species. Its numbers and distribution are currently increasing throughout Britain. Note that there is no evidence of this species ever being present outside of mainland Britain (on British islands, the Channel Islands, the Isle of Man or Ireland).

The current British population is found mainly in Scotland and is widespread throughout the River Tay catchment, and Perthshire, including a small population officially released as part of the Scottish Beaver Trial (SBT), Knapdale forest and mid-Argyll. There are also smaller numbers of free-living individuals in England, mostly in the south-west, including the River Otter in Devon (now part of the Devon Beaver trial). Several locations in England have recently been granted licences to have beavers within extensive enclosures. There are reports of sightings of beaver signs in parts of Wales, but it is unclear whether there are any established populations. The individuals used for the recent reintroductions are confirmed as being from Norway (SBT) or Bavaria (Tayside and Devon).

ECOLOGY

The Eurasian beaver is a semi-aquatic herbivore that can be found in fresh watercourses of varying size. It has been reported swimming in seawater but will not reside there. It will dam narrower watercourses to create deeper pools for breeding and feeding. Tree felling is more pronounced during the autumn/winter and when beavers are first colonising an area. The beaver's diet encompasses a wide range of riparian, emergent and aquatic vegetation. It is mainly nocturnal, though sightings during daylight hours are normal during spring/summer towards dusk and dawn.

IDENTIFICATION

Adult beavers are large (approx. 25 kg) with rotund bodies, and are larger than otters *Lutra lutra* and badgers *Meles meles*. The key identification features include size, orange incisors and broad, flat tails, which are dark grey in colouration and appear scale-covered. The beaver's fur is typically mid-brown in colour (variants are rare, but do exist across Europe, from black to light brown/blonde), although there are reports of small numbers of black individuals having been observed within the Tayside catchment.

BIBLIOGRAPHY

CAMPBELL-PALMER, R., GOW, D., NEEDHAM, R., JONES, S. & ROSELL, F. 2015. *The Eurasian Beaver*. Exeter: Pelagic Publishing and The Mammal Society.
SCOTTISH NATURAL HERITAGE. 2015. *Beavers in Scotland: A Report to Scottish Government*. Inverness: Scottish Natural Heritage.

AUTHOR   Roisin Campbell-Palmer

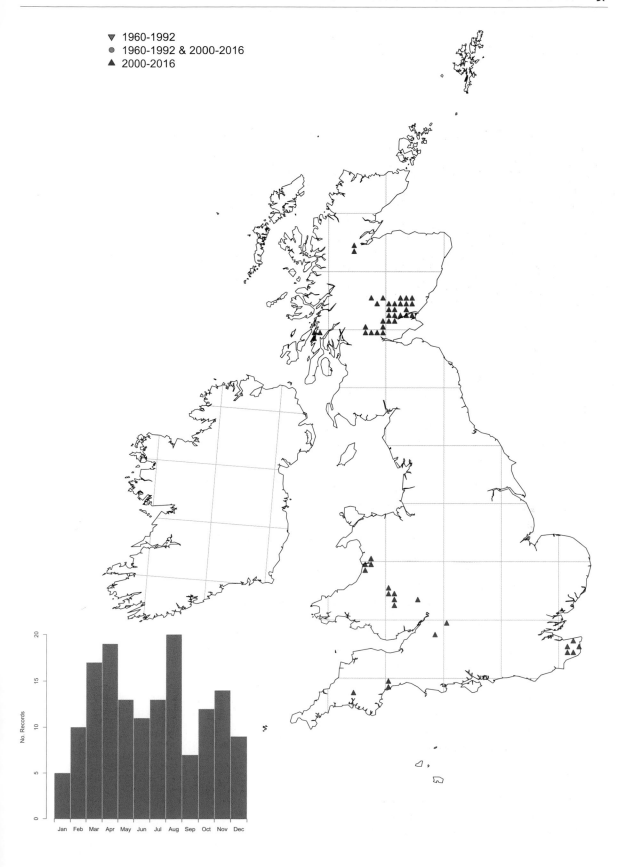

▼ 1960-1992
● 1960-1992 & 2000-2016
▲ 2000-2016

Hazel dormouse
Muscardinus avellanarius (LINNAEUS, 1758)

MATT BINSTEAD, BRITISH WILDLIFE CENTRE

DISTRIBUTION

The hazel dormouse is native to England and Wales, but not to Scotland or Ireland. A hundred years ago it was thought to be present in almost all English counties. Having become extinct from 12 northern and Midland English counties over the past century, it is now recorded predominantly in more southerly parts of England and along the English–Welsh border. The most northerly county where there is still a native population is Cumbria. It remains well distributed throughout Wales, with higher concentrations towards the east and south of the country. Overall, in Britain, the species' range is considered to be contracting. Data from the National Dormouse Monitoring Programme also indicates that the British population continues to decline. This is most likely the result of habitat fragmentation and inappropriate management.

ECOLOGY

The hazel dormouse is a small arboreal rodent closely associated with the early successional stages of woody vegetation, such as that found in traditional coppiced and managed woodland and scrub. The species is considered relatively sedentary and a poor disperser, and it is unlikely to occupy suitable local habitat in the absence of good linkages. It also lacks a caecum (a pouch which is connected to the small and large intestine), which severely restricts its ability to digest cellulose; hence, it is a sequential specialist feeder on pollen,

nectar, fruits and invertebrates. The species hibernates in nests on or near the ground between November and April. It also exhibits torpor when in its active period as a means to conserve energy.

IDENTIFICATION

The hazel dormouse is an agile arboreal rodent with large black eyes and prominent ears. The body length is 7–9 cm and its tail is well furred. Adults have sandy coloured coats on the back with paler bellies. Juveniles have more grey in the pelage. This species may be confused with the wood mouse *Apodemus sylvaticus* and bank vole, but its arboreal nature and furry tail are diagnostic. Hazelnuts are opened by dormice in a characteristic fashion, leaving a smooth opening, so feeding remains can be used to identify their presence.

BIBLIOGRAPHY

GOODWIN, C., HODGSON, D.J., AL-FULAIJ, N., BAILEY, S., LANGTON, S. & McDONALD, R.A. 2017. Voluntary recording scheme reveals ongoing decline in the United Kingdom hazel dormouse *Muscardinus avellanarius* population. *Mammal Review* 47 (3): 183–197.
WEMBRIDGE, D., AL-FULAIJ, N. & LANGTON, S. 2016. *The State of Britain's Dormice 2016*. Report for the People's Trust for Endangered Species. London: People's Trust for Endangered Species.

AUTHOR   Ian White

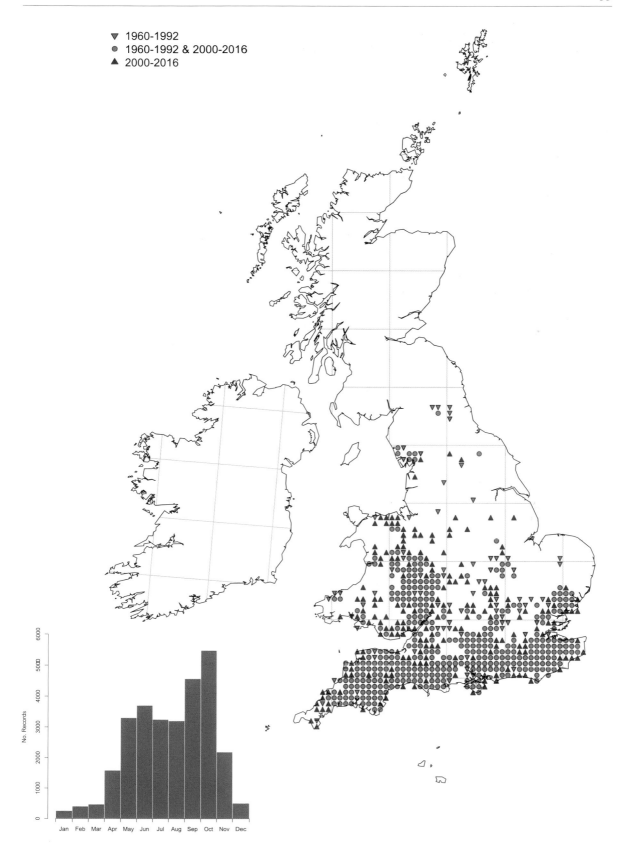

Bank vole

Myodes glareolus (SCHREBER, 1780)

GARY COX

DISTRIBUTION

The bank vole is present throughout mainland Britain and on the islands of Anglesey, Arran, Bute, Hayling, Handa, Ramsay, Scalpay, Ulva and the Isle of Wight, and with named subspecies on Mull, Raasay, Skomer and Jersey in the Channel Islands. It is not indigenous to Ireland, but was first recorded in County Kerry in 1964 (although probably accidentally introduced in the late 1920s) and is expanding its distribution, though it is not yet recorded in Northern Ireland. The bank vole is abundant throughout much of central and western mainland Europe.

ECOLOGY

The bank vole is common in broadleaved woodland, scrub, field margins and hedgerows but is also found in conifer plantations, fenland and road verges. It is active night and day, with peaks at dawn and dusk. Although mainly herbivorous, it consumes a wide range of foods including green leaves, dead leaves (winter), flowers, fruits, seeds, roots, mosses, fungi and invertebrates. The bank vole exhibits annual cycles in densities, with peaks in the autumn. Multi-annual cycles are known to occur in northern Europe.

IDENTIFICATION

The bank vole has a blunt nose, small ears, small eyes and a short tail. The back and top of the tail is chestnut brown, the underside is pale cream/grey. The head–body length is 8–12 cm. The tail is about half the length of the head and body. These features distinguish the bank vole from the field vole, which tends to be grey-brown in colour and has a tail about a third the length of the head and body. Characteristically, the bank vole feeds on hazelnuts, leaving a round hole with tooth marks across the inner edge but not on the surface of the nut. There are few other easy-to-spot field signs of its presence, although occasionally it may eat the bark of small trees.

BIBLIOGRAPHY

SHORE, R.F. & HARE, E.J. 2008. Bank vole *Myodes glareolus*. In S. Harris & D.W. Yalden (eds) *Mammals of the British Isles*, 4th edn. Southampton: The Mammal Society. pp. 88–99.

AUTHOR   John Gurnell

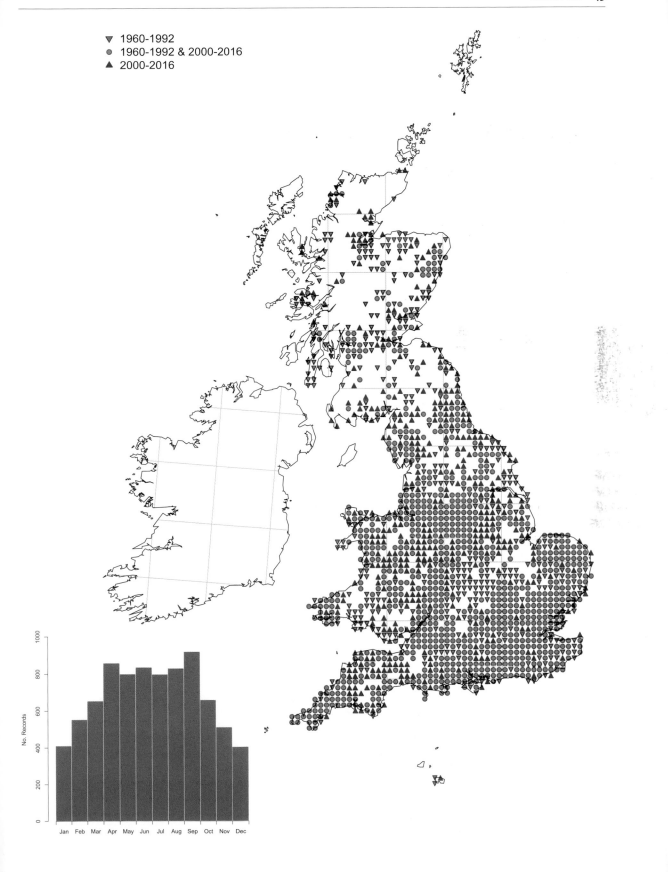

Field vole
Microtus agrestis (LINNAEUS, 1761)

DEREK CRAWLEY

DISTRIBUTION

The field vole occurs throughout mainland Britain and many offshore islands, but is absent from Lewis, Barra and some Inner Hebrides islands, Orkney, Shetland, Lundy, the Isles of Scilly and the Channel Islands. It is not found in Ireland but is common throughout mainland Europe.

ECOLOGY

The primary habitat of the field vole is rough, ungrazed grassland and young forest plantations. However, it may be found in woodland, hedgerows, bogs, dunes, moorland and road verges where grass is present. The field vole is mainly nocturnal, with activity peaks at dawn and dusk. It has a herbivorous diet, feeding on leaves and stems of grasses and occasionally mosses. At high densities, this species can cause damage to grassland and young plantations of broadleaved and fruit trees. It exhibits within-year and year-to-year changes in density, with some populations exhibiting three- to five-year multi-annual cycles.

IDENTIFICATION

The field vole has a blunt, rounded nose, small furry ears, small eyes and a short tail. The head–body length is 8–13 cm and its tail is about a third of the head–body length. Relative tail length and colouration are the main ways to distinguish it from the bank vole. Its nests are made out of woven grass, are approximately 10 cm in diameter, and may be found in the base of grass tussocks or in burrows. Signs of presence also include runways, and tunnels through long grass, with clipped grass and droppings inside.

BIBLIOGRAPHY

LAMBIN, X. 2008. Field vole *Microtus agrestis*. In S. Harris & D.W. Yalden (eds) *Mammals of the British Isles*, 4th edn. Southampton: The Mammal Society. pp. 100–107.

AUTHOR   John Gurnell

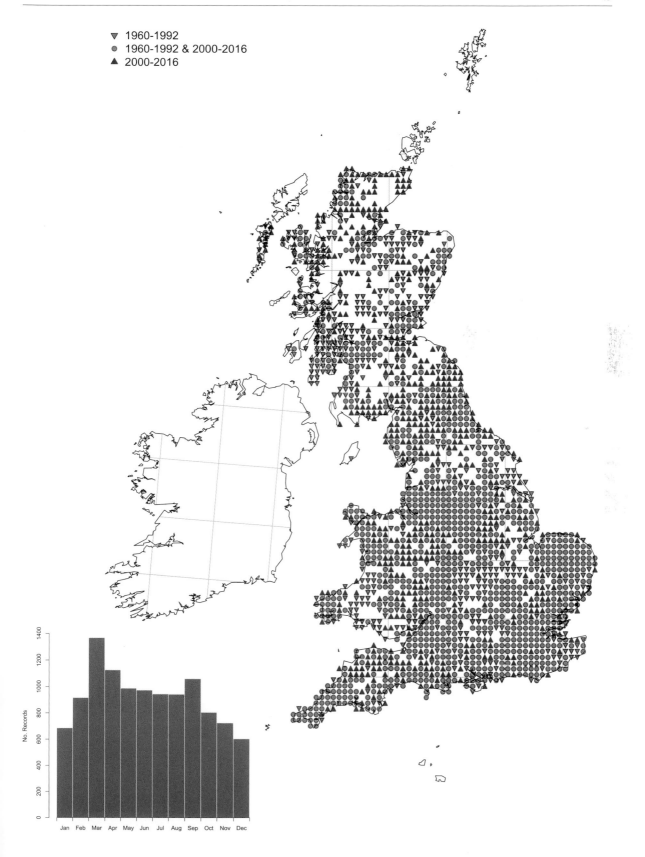

Common vole
Microtus arvalis (PALLAS, 1778)

SHANE AUSTIN

DISTRIBUTION

The common vole has a wide continental range from Spain to the Caucasus and central Siberia. In the British Isles, the species is restricted to Guernsey (*M. arvalis sarnius*) and to eight of the Orkney Islands (*M. arvalis orcadensis*): Mainland, Westray, Eday (introduced from Westray in 1987–88), Sanday, Burray, Hunda, South Ronaldsay and Rousay. The common vole was possibly present on Shapinsay until 1906. Neolithic remains from Holm of Westray might indicate an extinct population or transport of kills by raptors. Records of the species have been gathered from sightings, trapping, field signs and raptor pellets.

Recent comparative genetic studies of Orcadian and mainland European populations are consistent with a human introduction in the Neolithic period, at least 5,100 years BP (confirmed by radiocarbon dating), from coastal regions of what is now Belgium. It is not clear whether the Guernsey vole was introduced, or whether it colonised naturally before Guernsey became an island at the end of the last Ice Age.

ECOLOGY

The common vole in Orkney (known as the Orkney vole) is active day and night, showing short-term cycles of activity and rest with a period of roughly three hours. It is herbivorous, feeding on the fleshy parts of a wide variety of grasses. There has been no evidence of regular cycles in abundance. It is present in most habitats in Orkney, but largely absent from improved grasslands and arable crops. Agricultural areas – ditches, fence-lines and road verges – are important habitats. The Orkney vole is an important prey for hen harriers, short-eared owls and kestrels. In Guernsey, the common vole is more common in wet meadows than in drier areas.

IDENTIFICATION

Unmistakable since these are the only voles present on Guernsey and the Orkney islands.

BIBLIOGRAPHY

BOOTH, C. & BOOTH, J. 1994. *The Mammals of Orkney: A Status with an Appendix on Amphibians and Reptiles.* Kirkwall, Orkney: The Orcadian.

MARTÍNKOVÁ, N., BARNETT, R., CUCCHI, T., STRUCHEN, R., PASCAL, M., PASCAL, M., FISCHER, M.C., HIGHAM, T., BRACE, S., HO, S.Y., QUÉRÉ, J.P., O'HIGGINS, P., EXCOFFIER, L., HECKEL, G., HOELZEL, A.R., DOBNEY, K.M. & SEARLE, J.B. 2013. Divergent evolutionary processes associated with colonization of offshore islands. *Molecular Ecology* 22 (20): 5205–5220.

AUTHOR   Martyn Gorman

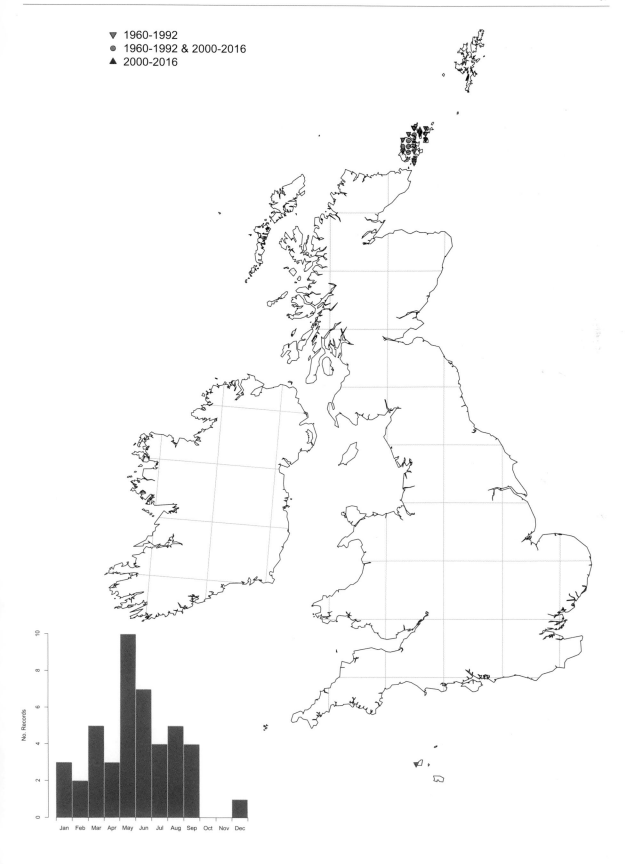

▼ 1960-1992
● 1960-1992 & 2000-2016
▲ 2000-2016

Water vole
Arvicola amphibius (LINNAEUS, 1758)

MATT BINSTEAD, BRITISH WILDLIFE CENTRE

DISTRIBUTION

In England, the water vole is widespread in the Midlands, East Anglia and southern England, the Isle of Wight and the Isle of Sheppey. It is patchily distributed in northern England, the Welsh Borders and the rest of Wales, with significant populations in Ceredigion, Monmouthshire and Anglesey. In Scotland, the Cairngorms, parts of Caithness and Sutherland, the north-west coast, the central belt (including within urban areas of Glasgow), Dumfries and Galloway, and the Scottish Borders have large water vole populations. The species has a very restricted distribution in Devon and is absent from Cornwall, the Isles of Scilly, the Isle of Man, the Channel Islands and the Scottish islands (with the exception of terrestrial/fossorial populations on small islands in the Sound of Jura).

The species has faced significant decline over the twentieth century as identified by a national survey in 1989–90. A repeat survey in 1996–97 showed a continuing contraction of range as well as a decline in numbers, attributable to predation from the American mink *Neovison vison* as well as to the loss and fragmentation of habitat. Reintroductions of the water vole have occurred in some areas.

Most records are from systematic field sign surveys (national, local or countywide, and proposed development sites) supplemented by reported sightings from the public.

ECOLOGY

The water vole is a herbivorous rodent, found in a wide variety of wetland habitats. It has a general preference for still or slow-flowing water, with grassy or herbaceous bankside vegetation, as well as emergent or in-channel vegetation. Terrestrial/fossorial populations are known in some areas of Scotland and parts of north-east England. It eats a variety of plant species during summer, and roots, bark, berries and stored food during winter. It is diurnal, with peak periods of activity around dawn and dusk.

IDENTIFICATION

The water vole is the largest species of vole in Britain, weighing 200–350 g. It has a rounded body, blunt muzzle and small ears, and the fur is chestnut brown. The head–body length is approximately 20 cm and the tail is approximately 60% of body length. A smaller melanistic (black) form occurs in parts of Scotland and has been recorded in parts of East Anglia. It can be confused with the larger brown rat *Rattus norvegicus*, but the latter has a more pointed muzzle and a pinkish scaly tail, whereas the tail of the water vole is brown and furry.

BIBLIOGRAPHY

STRACHAN, R. & JEFFERIES, D.J. 1993. *The Water Vole Arvicola terrestris in Britain 1989–90: Its Distribution and Changing Status.* London: The Vincent Wildlife Trust.

STRACHAN, C., STRACHAN, R. & JEFFERIES, D. 2000. *Preliminary Report on the Changes in the Water Vole Population of Britain as Shown by the National Surveys of 1989–1990 and 1996–1998.* London: The Vincent Wildlife Trust.

AUTHOR Mike Dean

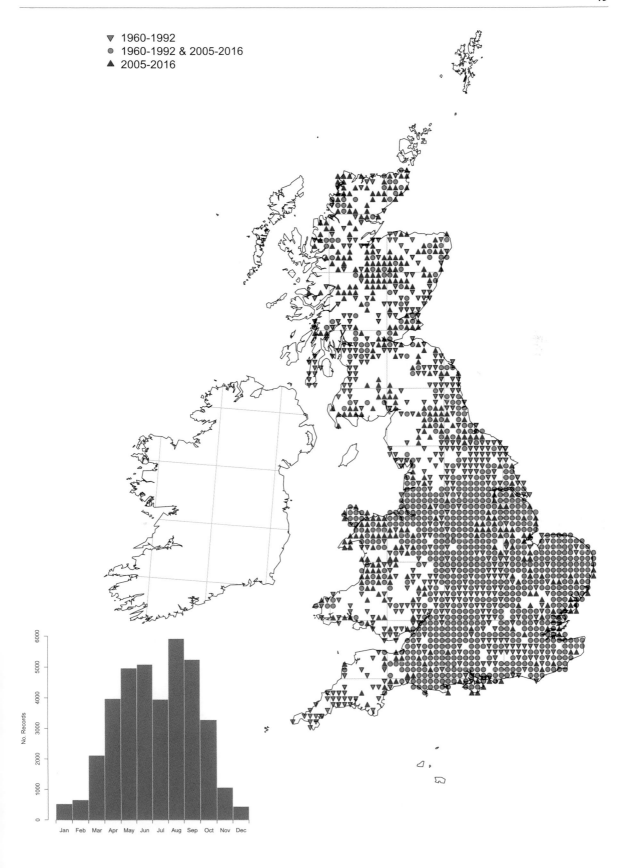

Harvest mouse

Micromys minutus (PALLAS, 1771)

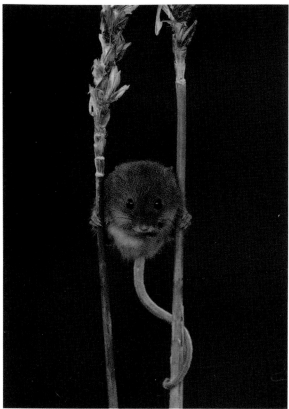

MATT BINSTEAD, BRITISH WILDLIFE CENTRE

DISTRIBUTION

Most records of harvest mouse populations lie to the south of a line drawn from south-east Wales to the North York Moors. The harvest mouse is also present in Pembrokeshire, north Wales, Cheshire, Anglesey and the Isle of Wight. Established populations are generally considered to be absent from Scotland, though there are a few records scattered across the country.

The harvest mouse is unobtrusive and is easily overlooked by inexperienced surveyors, so its current status is not fully understood. For example, surveys in Essex, Suffolk and Devon have found the harvest mouse to have a widespread distribution within these counties and this may be the case across much of its range. However, there has been a national decline in the availability of high-quality habitat for the species.

Most records are of the characteristic woven nest, which is easiest to find in autumn and winter. Occasionally, populations are estimated through findings in owl pellets, and harvest mice can be trapped in Longworth live traps, particularly in winter when they abandon the stalk zone and forage at ground level.

ECOLOGY

The harvest mouse is the smallest rodent in the UK. It has excellent climbing ability and occupies the stalk zone within a diverse range of grasses in summer and autumn. It is found in a wide range of habitats including rough grassland, reed beds, riparian margins, roadside verges, arable field margins and wild birdseed crops. The harvest mouse has a varied diet that includes seeds, berries and insects. Populations fluctuate greatly through the year with highest mortality in late winter.

IDENTIFICATION

The harvest mouse is a tiny mouse with golden-russet fur, pale belly and a semi-prehensile tail. Compared with other mice, the muzzle of a harvest mouse appears blunt and its eyes and ears are relatively smaller. The head–body length is 5–7 cm, and the tail is a similar length. The woven nest, constructed from split grass leaves that are still attached to their stalks, is a reliable field sign. These can be found as aerial nests in taller grasses, but in tussock-forming grasses they are usually hidden within the clump.

BIBLIOGRAPHY

DOBSON, J. 2012. The Harvest Mouse *Micromys minutus* Pallas, 1771 in Essex. *Essex Naturalist* 29: 129–135.

HARRIS, S. 1979. History, distribution, status and habitat requirements of the harvest mouse (*Micromys minutus*) in Britain. *Mammal Review* 9 (4): 159–171.

MEEK, M. & BULLION, S. 2012. Can the harvest mouse survive in a modern arable landscape? A Suffolk case study. *British Wildlife Magazine* 23 (6): 419–423.

AUTHOR   Simone Bullion

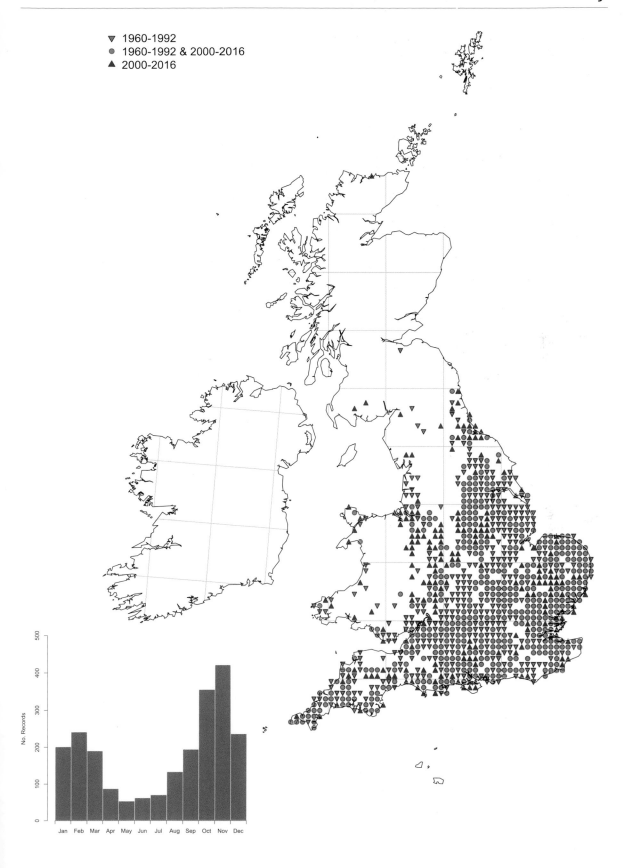

Wood mouse
Apodemus sylvaticus (LINNAEUS, 1758)

GARY COX

DISTRIBUTION

The wood mouse is ubiquitous throughout mainland Britain and Northern Ireland. It is also present on many islands such as Anglesey, the Isle of Man and the Isle of Wight, and on most small ones, including Fair Isle, Fetlar, Foula, Mainland, West Burra, Yell (Shetland); Copinsay, Eday, Hoy, Graemsay, Lings Holm, Mainland, North Ronaldsay, Sanday, Shapinsay, Stronsay (Orkney); Barra, Caenn Ear, Benbecula, Harris, Lewis, North Uist, St Kilda, South Uist (Outer Hebrides); Canna, Colonsay, Coll, Eigg, Gigha Iona, Jura, Lismore, Muck, Mull, Raasay, Rhum, Scarba, Shuna, Skye, Tiree, Islay (Inner Hebrides); Arran, Bute, Great Cumbrae (Clyde); Alderney, Guernsey, Herm, Jersey, Jethou, Sark (Channel Islands); St Marys, Tresco (Scilly Isles); Rathlin (Antrim); Tory Island (Donegal); Brownsea, Hayling Island, Hilbre, Skomer, Walney (misc.).

Records are derived from live-trapping studies of small mammals, pest control, cat kills and bird pellets. There is a scarcity of records from open mountainous areas, possibly reflecting a lack of survey effort, but the species has been recorded at 1,250 m in the Cairngorms and 1,000 m in County Kerry.

ECOLOGY

The wood mouse is present in most habitats, including woods, hedgerows, gardens, grassland and even on ploughed fields. Its diet consists of nuts, seeds, insects and green plant material. Breeding males have extensive ranges but size depends on habitat, whereas breeding female ranges are smaller and generally do not overlap with those of other breeding females. The wood mouse is primarily a nocturnal species, but pregnant females may also be diurnal. It is a good climber and an important prey species for many avian and mammalian predators.

IDENTIFICATION

The upper fur of the wood mouse is dark brown, while the underparts are white. There may be a yellow neck/chest spot or streak (occasionally spreading down and across the chest, almost joining the brown upper fur on neck, thus causing difficulties distinguishing from the yellow-necked mouse *Apodemus flavicollis*). However, melanistic and other colour forms of wood mouse can occur. The species has protruding eyes and ears. The head–body length is 8–11 cm, and the tail is just slightly shorter. Adult weights range from 14 g to 33 g (mean 20.4 g).

BIBLIOGRAPHY

FLOWERDEW, J.R. 2014. The study of British small mammals – 60 years on. *Mammal News* 169: 10–12.
FLOWERDEW, J.R. & TATTERSALL, F.H. 2008. Wood mouse *Apodemus sylvaticus*. In S. Harris & D.W. Yalden (eds) *Mammals of the British Isles*, 4th edn. Southampton: The Mammal Society. pp. 125–137.
GURNELL, J. & FLOWERDEW, J.R. 2006. *Live Trapping Small Mammals: A Practical Guide*, 4th edn. London: The Mammal Society.

AUTHOR   John Flowerdew

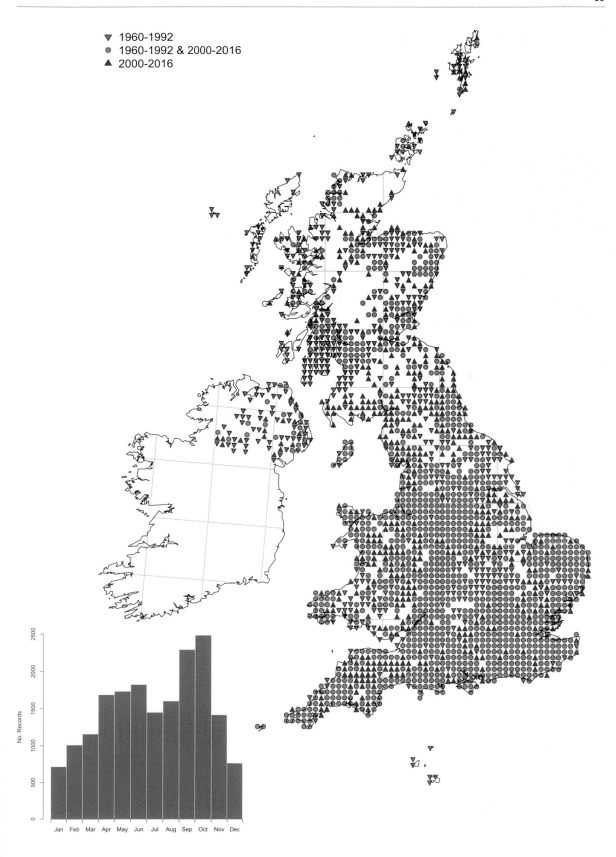

Yellow-necked mouse

Apodemus flavicollis (MELCHIOR, 1834)

MATT BINSTEAD, BRITISH WILDLIFE CENTRE

DISTRIBUTION

The yellow-necked mouse is widespread in east and south-east England and around the Welsh Marches, with only sporadic isolated records in south-west England, south-west Wales and north Norfolk. It is absent from Ireland, Scotland and northern England, with no records after the year 2000 being reported north of mid-Cheshire. Over 20 deciduous woodlands that were surveyed in northern and south-west England, west Wales and northern/central England failed to provide records in 1998. This species often occurs at low density, except for local 'hot spots' such as Gwent, and it is less common than the wood mouse in all but 15% of surveyed deciduous woodlands.

Historical and fossil evidence suggests a range extending north to Tyne and Wear/Northumberland. The yellow-necked mouse is absent from all islands, except for a recent record from Anglesey. Records of this species come from small mammal surveys using live-trapping, cat kills, bird pellets and pest control.

ECOLOGY

The yellow-necked mouse is present in mature deciduous woodland, hedgerows, gardens and orchard habitats, and it is also common in outbuildings. It is associated with ancient woodland, and with mature diverse nut/fruit-bearing trees with complex structure and much fallen timber. Its diet consists of fruits, seeds, green plant material and insects. It is a good arboreal climber with nocturnal habits. Low summer temperatures possibly limit its distribution due to their impact on tree seed masting and woodland diversity.

IDENTIFICATION

The yellow-necked mouse has dark brown upper fur with cream/white underparts. There is a distinctive yellow strip or 'collar' on the neck touching brown on either side, often extending centrally/posteriorly, which is equally clear in juveniles with grey fur. However, there is some potential confusion with the wood mouse, which also has yellow/buff markings. In the wood mouse these vary from being completely absent, to being a spot or star, to being an anterior–posterior stripe. The yellow-necked mouse has protruding eyes and ears similar to the wood mouse. The head–body length is 9–13 cm and the tail is slightly longer than this. Adults have a body mass 1.5 times that of an adult wood mouse, reaching 45 g or more (range 20–50 g). This species is more likely than the wood mouse to emit a high-pitched 'scream' when it is held.

BIBLIOGRAPHY

MARSH, A.C.W. & MONTGOMERY, W.I. 2008. Yellow-necked mouse *Apodemus flavicollis*. In S. Harris & D.W. Yalden (eds) *Mammals of the British Isles*, 4th edn. Southampton: The Mammal Society. pp. 137–141.

MARSH, A.C.W., POULTON, S. & HARRIS, S. 2001. The yellow-necked mouse *Apodemus flavicollis* in Britain: status and analysis of factors affecting distribution. *Mammal Review* 31 (3–4): 203–227.

MONTGOMERY, W.I. 1985. Interspecific competition and comparative ecology of two congeneric species of mice. In L.M. Cook (ed.) *Case Studies in Population Biology*. Manchester: Manchester University Press. pp. 126–187.

AUTHOR   John Flowerdew

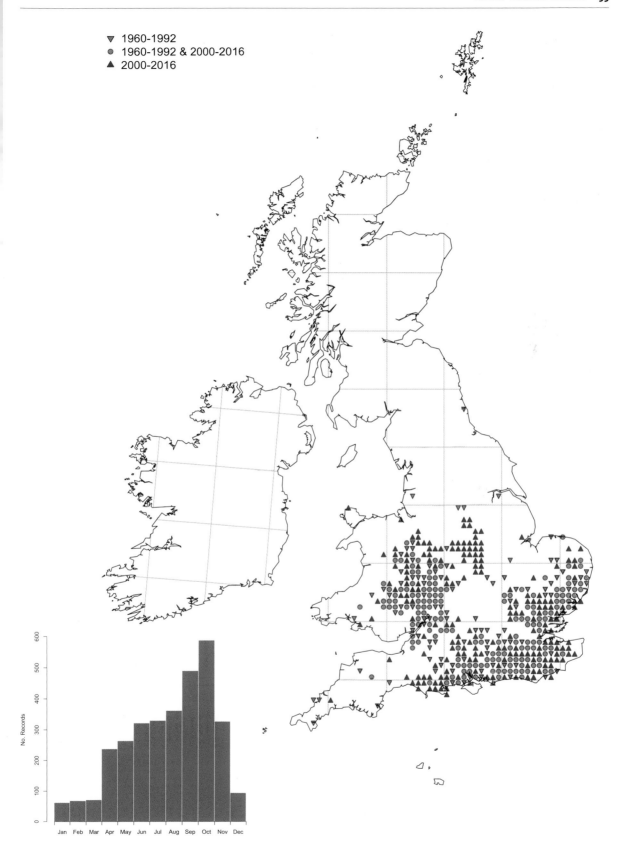

Brown rat
Rattus norvegicus (BERKENHOUT, 1769)

MALCOLM WELCH

DISTRIBUTION

The brown rat is believed to have originated in central Asia but now has a worldwide distribution. It spread across Europe in the eighteenth century, and has been present in Great Britain and Northern Ireland since the 1720s. The brown rat is found throughout Britain except in some mountain regions in Scotland and on some smaller offshore islands, but it is highly under-recorded.

ECOLOGY

The brown rat is mainly nocturnal, omnivorous, and lives in colonies consisting of several smaller social groups. It is a commensal rodent but can be found in some places away from humans, such as salt marshes in coastal areas and along watercourses. The brown rat occupies a wide range of urban and agricultural habitats: buildings, warehouses, farms, barns, waterways, sewers, crop fields and field margins – especially areas with good ground cover near water. When the brown rat arrived in Britain, it largely displaced the black rat. It is regarded as a pest because it eats and contaminates stored food, damages buildings and wiring, fouls surfaces, and acts as a vector of disease.

IDENTIFICATION

Larger than the black rat, the brown rat has a blunter nose, smaller eyes, and smaller, hair-covered ears. The head–body length is 15–27 cm, and its almost hairless pinkish tail is shorter than its body. Body colours range from brown to black, but its underparts are paler. Its presence can be detected by droppings, smear marks, footprints, trails and holes, but these may not be easy to distinguish from those of the black rat. The brown rat can swim well and can be confused with the water vole, which has a shorter, blunter nose, smaller ears and a shorter, hairy tail that is half the length of its body. However, the body of the water vole is more buoyant in the water.

BIBLIOGRAPHY

QUY, R.J. & MACDONALD, D.W. 2008. Common rat *Rattus norvegicus*. In S. Harris & D.W. Yalden (eds) *Mammals of the British Isles*, 4th edn. Southampton: The Mammal Society. pp. 149–155.

AUTHOR   John Gurnell

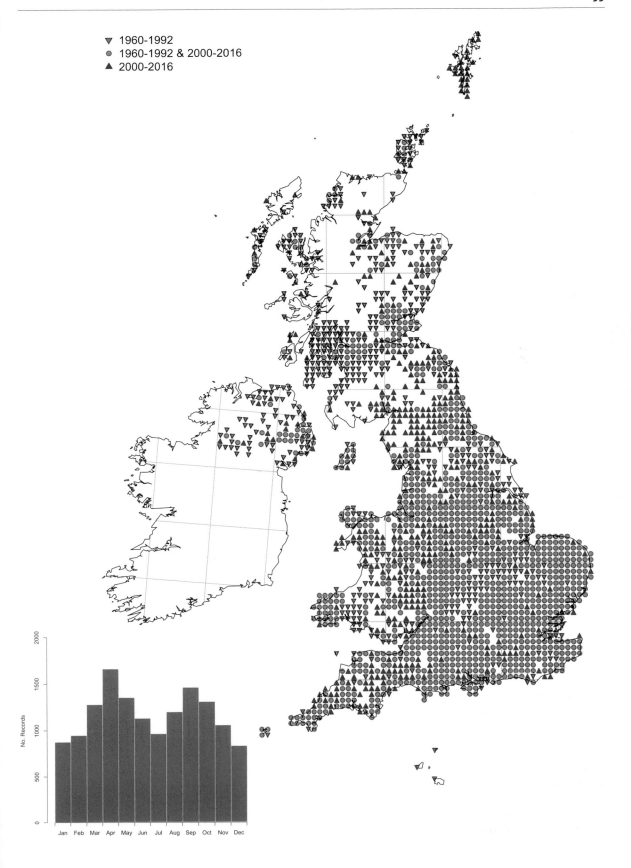

Black rat

Rattus rattus (LINNAEUS, 1758)

MATT BINSTEAD, BRITISH WILDLIFE CENTRE

DISTRIBUTION

The black rat is thought to have originated in South-East Asia and has been spread to many parts of the world by seafaring humans since the post-glacial period. After being introduced to Britain, probably with the Romans (first century BC to fifth century AD), it became widespread in Britain and across Europe. However, since the eighteenth century it has been almost entirely replaced by the brown rat. On the mainland, there are occasional records in and around dockyards or harbours, accidentally introduced in ships' merchandise (e.g. Port of Tilbury and other London docks). Most former island colonies (e.g. Lundy, Shiant Isles) have now been eradicated. Overall there has been a dramatic decline (99%) in the distribution of this species between the historical and current atlas period: there were 82 positive hectads between 1960 and 1992, 16 between 2000 and 2009 (although some records may be caused by misidentifications) and one (the Shiants, which has subsequently been eliminated) between 2010 and 2016. Nevertheless the species is not yet classified as functionally extinct in the UK owing to the lack of systematic surveys.

ECOLOGY

A commensal rodent, on the mainland it is almost always found in and around buildings (e.g. warehouses, restaurants, supermarkets, department stores), but can live among rocks and cliffs on islands. The black rat is mostly nocturnal, omnivorous, and lives in social groups each dominated by a male. Black rats are agile climbers and are often found in lofts or roof areas. Just like the brown rat, the black rat is regarded as a pest. On islands, it can have serious impacts on seabird colonies and endemic species.

IDENTIFICATION

The black rat is similar in appearance to the brown rat, but it is slightly smaller with a more slender body, larger eyes and ears, and a longer tail. The head–body length is 15–24 cm. Coat colours vary from brown to black, although three different colour morphs have been described. Droppings, smear marks, footprints, trails and holes are signs of presence, but these may not be easy to distinguish from the brown rat.

BIBLIOGRAPHY

TWIGG, G.I., BUCKLE, A.P. & BULLOCK, D.J. 2008. Ship rat *Rattus rattus*. In S. Harris & D.W. Yalden (eds) *Mammals of the British Isles*, 4th edn. Southampton: The Mammal Society. pp. 155–158.

AUTHOR   John Gurnell

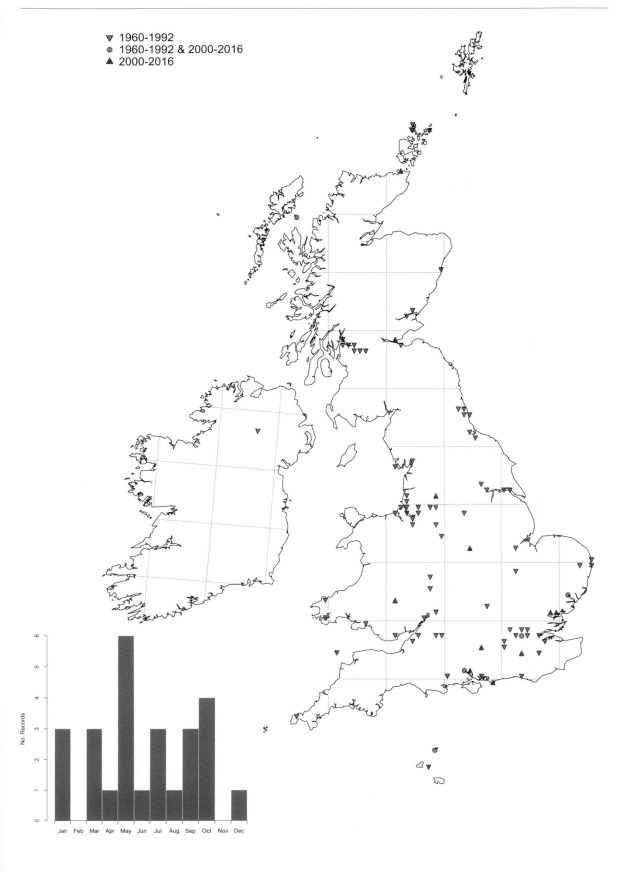

Wildcat
Felis silvestris (MILLER, 1907)

WENDY ELLISON

DISTRIBUTION

Although once found across Britain, the current distribution of the wildcat is restricted to Scotland. Systematic surveys in 1983–87 and 2006–08 have reported populations in Caithness and Sutherland, Easter Ross, the Cairngorms, Moray and Aberdeenshire, Morvern, Ardnamurchan and Argyll. Since 2013, no reliable records have been reported in the northern Highlands and very few records in the west outside Morvern, Ardnamurchan and Kintyre.

The wildcat is extremely rare with recent population estimates indicating very low densities (0.03–0.99 per km²). Recent declines may have been driven by a number of factors including rabbit population declines, a series of unusually hard winters in 2009–11, hybridisation with domestic cats and persecution. Currently, as part of the national action plan, *in situ* conservation efforts are focused on five priority areas, each in the range of 200–500 km². As part of the same action plan, there is also a conservation breeding programme for future release to supplement populations in the wild.

Records are obtained from questionnaires and public sightings, but mainly from camera-trap footage.

ECOLOGY

The wildcat in Scotland is typically a species of woodland edge or scrub and adjacent rough grasslands, riparian habitats and moorland fringes. Depending on the landscape, individuals may move seasonally between upland areas in the spring and summer to lower forest habitat over winter. Home range size can vary from 1 km² to 27 km² depending on prey availability and time of year: males range further in the late winter breeding season. The species is largely solitary except when breeding. It tends to be more active at night and around dusk and dawn, but can also be active during the day. The diet varies markedly across its distribution: there is a preference for rabbits where they are available, and elsewhere mice and voles are primary prey items.

IDENTIFICATION

The wildcat displays a striped tabby-coloured coat pattern, with a thick, blunt, black-tipped tail with clear black bands and a dorsal stripe down the back that stops at the base of the tail. It also has distinctive stripes on the nape (four thick stripes) and shoulder (two thick stripes). On average, it is a little larger than the domestic cat, although there is considerable overlap in size. The wildcat generally has a broader face than the domestic cat.

Hybridisation with the domestic cat is a major threat to the wildcat in Scotland, with most, if not all, remaining wildcats having at least some domestic cat ancestry. Hybrids can show a range of intermediate coat characteristics. Genetic tests and coat marking are both indicative of the level of hybridisation and both are correlated.

BIBLIOGRAPHY

DAVIS, A.R. & GRAY, D. 2010. *The Distribution of Scottish Wildcats (Felis silvestris) in Scotland (2006–2008)*. Scottish Natural Heritage Commissioned Report No. 360. Inverness: Scottish Natural Heritage.

EASTERBEE, N., HEPBURN, L.V. & JEFFERIES, D.J. 1991. *Survey of the Status and Distribution of the Wildcat in Scotland, 1983–1987*. Edinburgh: Nature Conservancy Council for Scotland.

KITCHENER, A.C., YAMAGUCHI, N., WARD, J.M. & MACDONALD, D.W. 2005. A diagnosis for the Scottish wildcat (*Felis silvestris*): a tool for conservation action for a critically-endangered felid. *Animal Conservation* 8 (3): 223–237.

AUTHORS   Ruairidh Campbell and Jenny Bryce

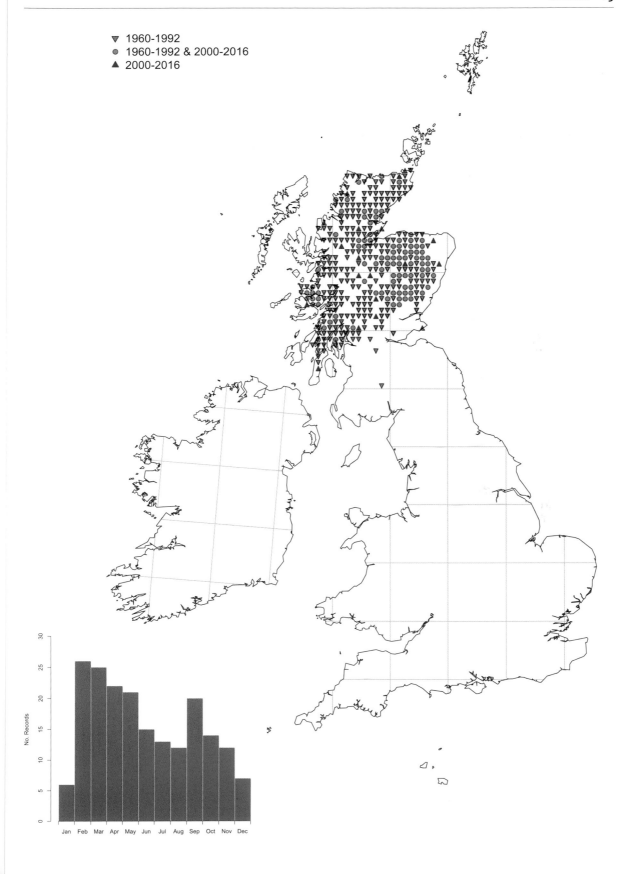

▽ 1960-1992
● 1960-1992 & 2000-2016
▲ 2000-2016

Badger
Meles meles (LINNAEUS, 1758)

PADRAIG KAVANAGH

DISTRIBUTION

The badger is widely distributed in all counties of England, Scotland, Wales and Northern Ireland, having recovered from a decline prior to the early 1900s. It is also present on some offshore islands, including Anglesey, Arran, Canvey, the Isle of Wight, the Isle of Sheppey and Skye. Badger density is greatest in the south and west of England and Wales, and lowest in the north of Scotland. In Northern Ireland, it appears to be evenly distributed across the country. During the most recent national survey of badgers in Northern Ireland, 75% of the surveyed 1 km² grids contained at least one sett. In England and Wales, over 50% contained an 'active' sett of any type, and in Scotland, only 15% contained a sett of any type.

ECOLOGY

The badger is a terrestrial carnivore found in a wide range of rural and urban habitats, with a preference in rural areas for mixed farmland (for foraging and cover) and well-drained soils (for setts). Omnivorous and opportunistic, the badger feeds on a wide range of invertebrates (earthworms, beetle larvae), vertebrate prey (small mammals, carrion), fruit (blackberries, elder), roots and cereals. It is nocturnal.

IDENTIFICATION

The badger is a large, well-built animal with a medium-length muzzle, short legs and tail. The face is white with black stripes from ears to muzzle; the back and sides appear grey and the underside is usually dark. The badger is unlikely to be confused with any other mammal.

BIBLIOGRAPHY

DELAHAY, R., WILSON, G., HARRIS, S. & MACDONALD, D.W., 2008. Badger *Meles meles*. In S. Harris & D.W. Yalden (eds) *Mammals of the British Isles*, 4th edn. Southampton: The Mammal Society. pp. 425–436.

JUDGE, J., WILSON, G.J., MACARTHUR, R., DELAHAY, R.J. & McDONALD, R.A. 2014. Density and abundance of badger social groups in England and Wales in 2011–2013. *Scientific Reports* 4: 3809.

RAINEY, E., BUTLER, A., BIERMAN, S. & ROBERTS, A.M.I. 2009. *Scottish Badger Distribution Survey 2006–2009: Estimating the Distribution and Density of Badger Main Setts in Scotland*. Innerleithen: Scottish Badgers and Biomathematics and Statistics Scotland.

REID, N., ETHERINGTON, T.R., WILSON, G., McDONALD, R.A. & MONTGOMERY, W.I. 2008. *Badger Survey of Northern Ireland 2007/08*. Report prepared by Quercus and Central Science Laboratory for the Department of Agriculture & Rural Development (DARD), Northern Ireland, UK. Belfast: Department of Agriculture & Rural Development.

SMAL, C. 1995. *The Badger and Habitat Survey of Ireland*. Dublin: Government Stationery Office.

AUTHOR   Penny Lewns

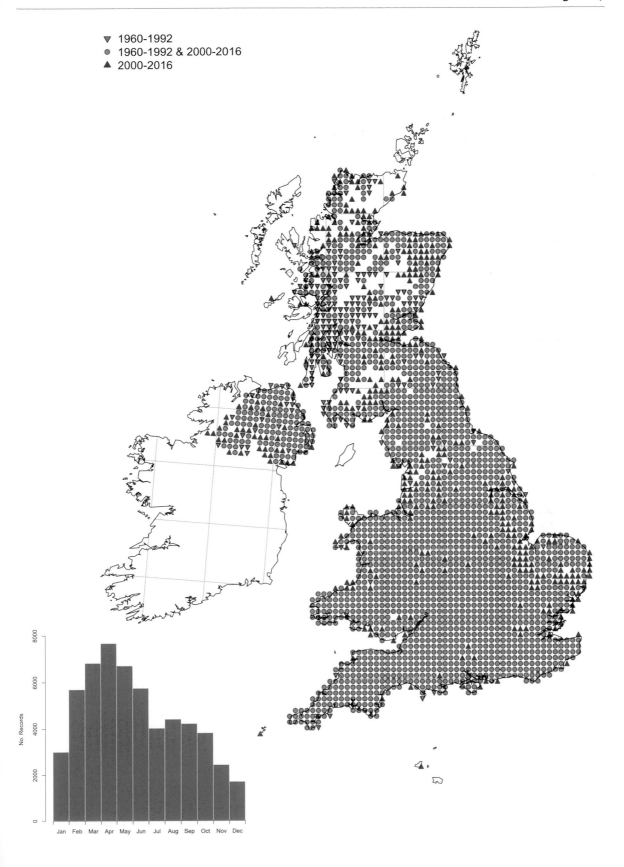

Otter

Lutra lutra (LINNAEUS, 1758)

MARK BALDWIN

DISTRIBUTION

The otter is present more or less throughout the UK, having recovered from a catastrophic British decline during the 1950s to the 1970s. Systematic national surveys have been carried out a number of times since 1977. The latest national surveys show it is still recovering in parts of England with few records from the south-east and parts of the Peak District, Cheshire and Lancashire. It was found at more than 75% of survey sites in all catchment areas in Wales other than Anglesey (68%) and Mid Glamorgan (71%). The otter population is stable in many parts of Scotland. Recovery of populations continues in Strathclyde, Ayrshire, Forth and the Borders, but more than 80% of sites were occupied in all counties in the latest survey. It is present throughout the whole of Ireland and is recorded on all Scottish islands of any size including the Orkneys and Shetland Isles. The otter is absent from the Isle of Man, the Isles of Scilly, Lundy and the Channel Islands.

Most records are from systematic sign surveys supplemented by systematic collection of road casualties.

ECOLOGY

The otter is a semi-aquatic carnivore found on inland watercourses, including still water, and also along the coast. It feeds mainly on fish, amphibians and crayfish. In freshwater habitats it is mainly nocturnal, but is more diurnal when living on the coast.

IDENTIFICATION

The otter is a long-bodied animal, much larger than a cat, with a stout tapering tail. It is mid- to dark brown in colour. It may be confused with the American mink, but the latter is much smaller, very dark brown in colour, and has a fluffy tail.

BIBLIOGRAPHY

CRAWFORD, A. 2010. *Fifth Otter Survey of England 2009–2010*. Technical report. Bristol: Environment Agency.

REID, N., HAYDEN, B., LUNDY, M.G., PIETRAVALLE, S., McDONALD, R.A. & MONTGOMERY, W.I. 2013. *National Otter Survey of Ireland 2010/12*. Irish Wildlife Manuals No. 76. Dublin: National Parks and Wildlife Service, Department of Arts, Heritage and the Gaeltacht.

STRACHAN, R. 2007. *National Survey of Otter Lutra lutra Distribution in Scotland 2003–04*. Scottish Natural Heritage Commissioned Report No. 211 (ROAME No. F03AC309). Inverness: Scottish Natural Heritage.

STRACHAN, R. 2015. *Otter Survey of Wales 2009–10*. Cardiff: Natural Resources Wales.

AUTHOR   Paul Chanin

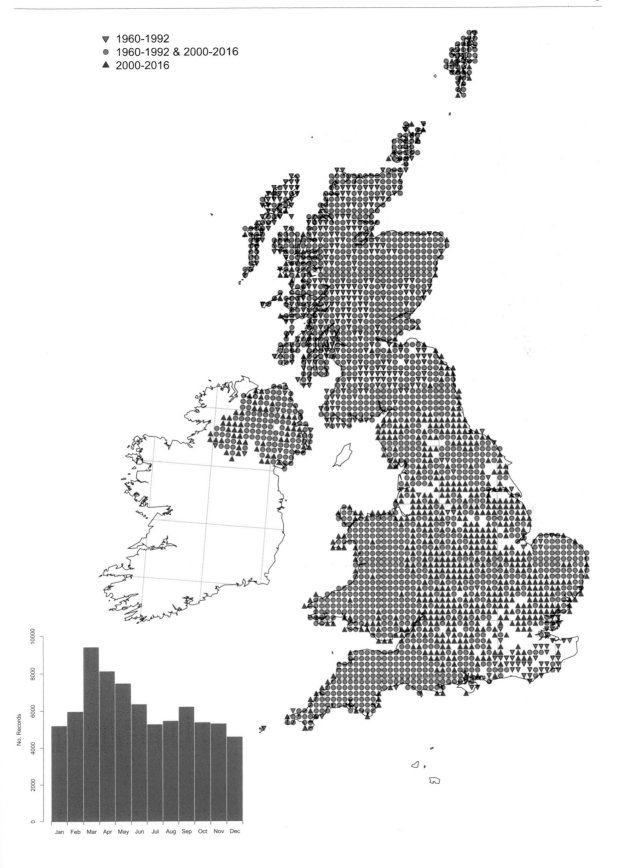

1960-1992
1960-1992 & 2000-2016
2000-2016

Pine marten
Martes martes (LINNAEUS, 1758)

PADRAIG KAVANAGH

DISTRIBUTION

The pine marten has expanded its range in Scotland significantly over the past few decades, following an extensive population decline during the nineteenth century. It is now widespread in northern and central Scotland north of the industrialised central belt. It is also present on the islands of Skye and Mull. There are isolated populations in southern Scotland, in Dumfriesshire and the Scottish borders, originating from reintroductions. It has a scattered distribution in Northern Ireland.

The pine marten remains absent from most of England where the population has not recovered from its historical decline. In 2015, the presence of a small number of pine martens was confirmed in Shropshire, and in 2016 footage of a pine marten was recorded in the New Forest. The origins of these animals is currently unclear though they may derive from covert releases. Elsewhere, there are occasional records from Cumbria, Northumberland, North Yorkshire and the Peak District.

A pine marten population reinforcement was carried out in Wales in 2015 and 2016 by translocating pine martens from Scotland. A population is now established in mid-Wales in parts of Ceredigion and Powys, with occasional records from Carmarthenshire and Snowdonia.

Most records from recent distribution surveys in Scotland are derived from scats, verified by DNA analysis, collected from systematic transect surveys. Records are also supplemented by road casualties, live sightings, and photos and videos from camera traps.

ECOLOGY

The pine marten favours three-dimensional habitats, preferentially woodland, and dens in tree cavities, branches, wind-blown trees, squirrel dreys and bird nests. It has a varied omnivorous diet, principally comprising small mammals, fruits, birds and invertebrates. Pine martens are mostly nocturnal and crepuscular. They are solitary and mate during July–August with 1–5 young (kits) born the following March–April.

IDENTIFICATION

The pine marten is the size of a small domestic cat. It has a slim body with brown fur and a distinctive cream 'bib' on the throat and chest, which can be used to identify individuals. The pine marten has a long bushy tail, prominent rounded ears and relatively long legs.

BIBLIOGRAPHY

BIRKS, J.D.S. & MESSENGER, J. 2010. *Evidence of Pine Martens in England and Wales 1996–2007: Analysis of Reported Sightings and Foundations for the Future.* Ledbury: The Vincent Wildlife Trust.

CROOSE, E., BIRKS, J.D.S. & SCHOFIELD, H.W. 2013. *Expansion Zone Survey of Pine Marten (Martes martes) Distribution in Scotland.* Scottish Natural Heritage Commissioned Report No. 520. Ledbury: Scottish Natural Heritage.

CROOSE, E., BIRKS, J.D.S., SCHOFIELD, H.W. & O'REILLY, C. 2014. *Distribution of the Pine Marten (Martes martes) in southern Scotland in 2013.* Scottish Natural Heritage Commissioned Report No. 740. Ledbury: Scottish Natural Heritage.

AUTHOR Elizabeth Croose

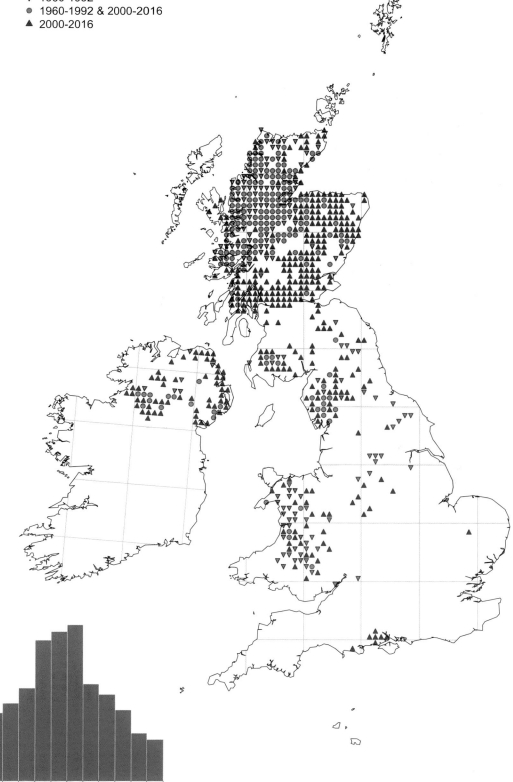

▼ 1960-1992
● 1960-1992 & 2000-2016
▲ 2000-2016

American mink
Neovison vison (SCHREBER, 1777)

DEREK CRAWLEY

DISTRIBUTION

The American mink is native to North America, but is now widespread in Britain and Ireland following escapes and releases from mink farms since the mid-twentieth century. The establishment, expansion and coalescence of feral populations was facilitated by the multiple locations of mink farms, and the initial scarcity of native competitors – the otter and the polecat (the latter is not native to Ireland). Now, the American mink is widespread and present in most counties, though it is scarce or absent in some areas such as northern Scotland. The species is present on some offshore islands such as Mull, Skye and Arran.

Since the 1970s, field signs of American mink have been routinely recorded during national surveys of otters in Britain and Ireland (and, more recently, of water voles in Britain). These have provided evidence of changes in American mink distribution and, possibly, abundance. In Britain, post-2000 surveys have generally recorded a decline in the number of sites positive for American mink. For example, the fifth otter survey of England found that between 2000–02 and 2009–10 the number of survey sites positive for American mink had declined in ten out of twelve catchment areas, with six catchments recording declines of more than 50%. Naturalists also report declines in American mink abundance in some areas such as the Midlands and south-west England, and there are co-ordinated and effective control efforts over large areas of Scotland.

ECOLOGY

The American mink is a semi-aquatic generalist carnivore found on, or close to, a wide variety of freshwater bodies and some coastal habitats (e.g. sheltered, gently shelving rocky coastlines). It feeds on a wide range of prey including birds, mammals, fish, amphibians and crustaceans. Recent research suggests that the American mink in England has become more diurnal in its activity where it faces competition with otters and/or polecats.

IDENTIFICATION

The American mink is a medium-sized slim-bodied mustelid, very dark brown in colour with a fluffy tail. It is much smaller than the otter and more buoyant when swimming. Paler coat colours occur occasionally and the species often has a pale patch under the chin.

BIBLIOGRAPHY

CRAWFORD, A. 2010. *Fifth Otter Survey of England 2009–2010*. Technical report. Bristol: Environment Agency.

McDONALD, R.A., O'HARA, K. & MORRISH, D.J. 2007. Decline of invasive alien mink (*Mustela vison*) is concurrent with recovery of native otters (*Lutra lutra*). *Diversity and Distributions* 13 (1): 92–98.

AUTHOR   Johnny Birks

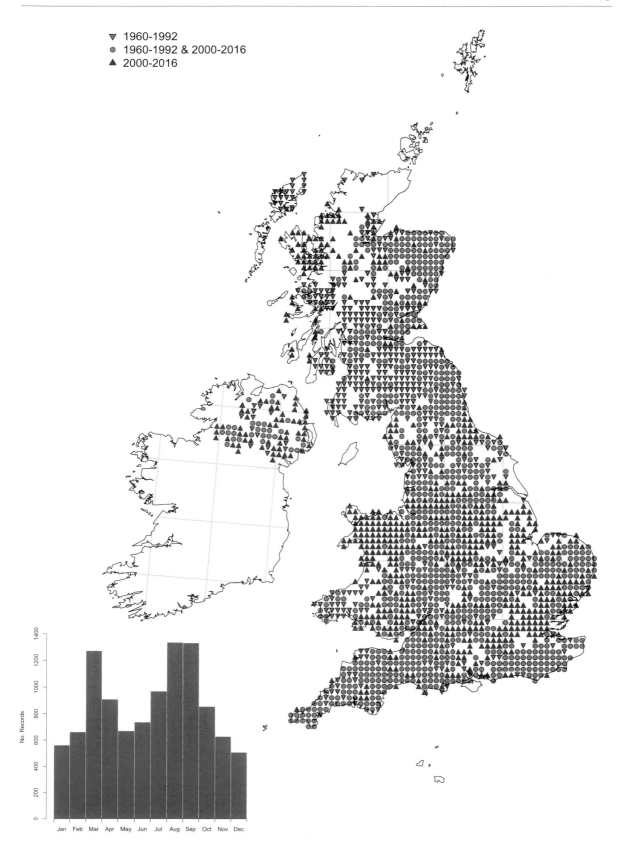

- ▽ 1960-1992
- ● 1960-1992 & 2000-2016
- ▲ 2000-2016

Red deer
Cervus elaphus (LINNAEUS, 1758)

LIZ BRACKEN

DISTRIBUTION

The red deer is common in the Scottish Highlands and Islands, Dumfriesshire, the Lake District, East Anglia and the south-west of England. It is also present in parts of Northern Ireland. Further populations, largely descended from park escapes and deliberate releases, are found in the north of England, north and west Midlands, East Anglia, the New Forest, Sussex, Surrey and limited areas of Wales. Records are derived from the British Deer Society five-yearly Deer Distribution Survey (2016 data), together with other sightings, camera-trap images and road casualties.

ECOLOGY

A predominantly grazing herbivore that prefers forest and woodland habitats, the red deer has adapted successfully to open moors and hills. Behaviour can vary between habitats: on open ground, large herds can form, while in woodland, deer tend to live in smaller groups or singly. In Scotland, animals tend to spend the days on the open hills and descend to lower ground at night. The sexes tend to live apart for most of the year and only come together for the rut, which takes place from the end of September to November. A single calf is born between mid-May and July; twins are very rare.

In harsher, upland habitats, or when overpopulated, females (hinds) may not breed until 3–4 years old and mature hinds may not ovulate when lactating.

IDENTIFICATION

The largest British land mammal, a red deer stag can stand as high as 140 cm at the shoulder; hinds are typically at least a third smaller and lighter. Different climatic and environmental conditions lead to wide variations in species size. Southern, lowland and woodland deer tend to be notably larger than those living in the north in open environments.

The summer coat is a reddish brown, turning darker and greyer in winter, with a pale caudal patch that extends above a relatively short tail. Most mature stags develop a shaggy mane as the rut approaches and carry wide, branched antlers. In wild animals, antlers may have up to 18 points (tines), including two brow tines, and sometimes more, especially in lowland or park specimens.

BIBLIOGRAPHY

BRITISH DEER SOCIETY. 2017. 2016 deer distribution survey. Accessed at https://www.bds.org.uk/index.php/news-events/297-2016-deer-distribution-survey-results-now-available (30 July 2019).

PARLIAMENTARY OFFICE OF SCIENCE AND TECHNOLOGY. 2009. Wild deer. POSTnote No. 325. Accessed at http://researchbriefings.parliament.uk/ResearchBriefing/Summary/POST-PN-325 (30 July 2019).

AUTHORS Hugh Rose and Charles Smith-Jones

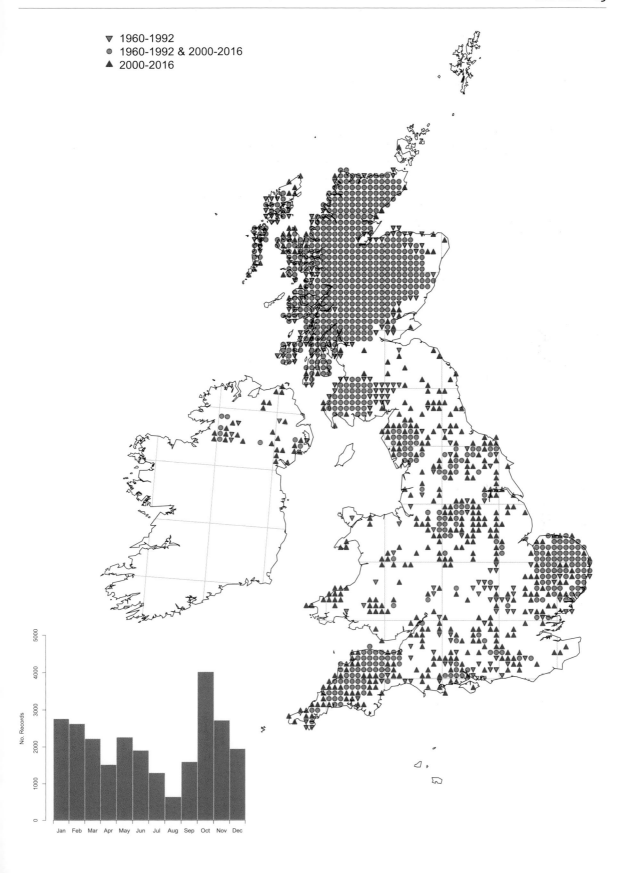

1960-1992
1960-1992 & 2000-2016
2000-2016

Sika deer

Cervus nippon (TEMMINCK, 1838)

SIMON JOHNSON

DISTRIBUTION

The sika deer is native to Japan, where the species is considered abundant and increasing. Deliberate and accidental introduction into southern and northern England and into Scotland occurred from 1870 onwards. The species is now widespread and is expanding in Scotland, particularly in the west, while patchy populations are present across England, with locally strong presences in Dorset, Lancashire and Cumbria. The species is also present in much of Northern Ireland, particularly County Fermanagh and County Tyrone. Its presence in Wales is restricted and extremely localised. Records are derived from the British Deer Society five-yearly Deer Distribution Survey (2016 data), together with other sightings, camera-trap images and road casualties.

ECOLOGY

The sika deer is a browsing and grazing herbivore that takes a wide variety of food plants. It prefers coniferous woodlands and heathlands associated with acid soils, though other habitats are used as populations expand. The sexes tend to form separate herds for much of the year, coming together only for the rut, and are frequently encountered singly or in small groups at other times. Where they are locally numerous, large herds can form. Most activity is nocturnal or very close to the hours of darkness. The rut takes place between late September and November, and a single calf is born in the following May or June.

IDENTIFICATION

Similar in size to the fallow *Dama dama*, the sika deer stands about 100 cm high at the shoulder though males (stags) are significantly larger. The summer coat is comparable to that of the fallow deer, being chestnut brown with creamy spots and a black dorsal stripe, turning a darker grey/brown with less apparent spots in winter. The tail is shorter than that of the fallow deer and has a less distinct black stripe down the centre; the white rump patch is bordered by an indistinct black edge, which extends to the hocks. There is normally a distinct pale metatarsal gland marking on the hind legs. Males have branched antlers similar to those of red deer, but typically with no more than eight tines (points) and with a single brow tine leaving the main beam at an acute angle.

The sika and red deer are capable of hybridisation. The first cross between the species has the appearance of both parents, but subsequent backcrossing makes many hybrids hard to detect in the field.

BIBLIOGRAPHY

BRITISH DEER SOCIETY. 2017. 2016 deer distribution survey. Accessed at https://www.bds.org.uk/index.php/news-events/297-2016-deer-distribution-survey-results-now-available (30 July 2019).

HARRIS, R.B. 2015. *Cervus nippon. The IUCN Red List of Threatened Species* 2015: e.T41788A22155877. Accessed at http://dx.doi.org/10.2305/IUCN.UK.2015-2.RLTS.T41788A22155877.en (30 July 2019).

PARLIAMENTARY OFFICE OF SCIENCE AND TECHNOLOGY. 2009. Wild deer. POSTnote No. 325. Accessed at http://researchbriefings.parliament.uk/ResearchBriefing/Summary/POST-PN-325 (30 July 2019).

AUTHORS   Rory Putman, Josephine Pemberton and Charles Smith-Jones

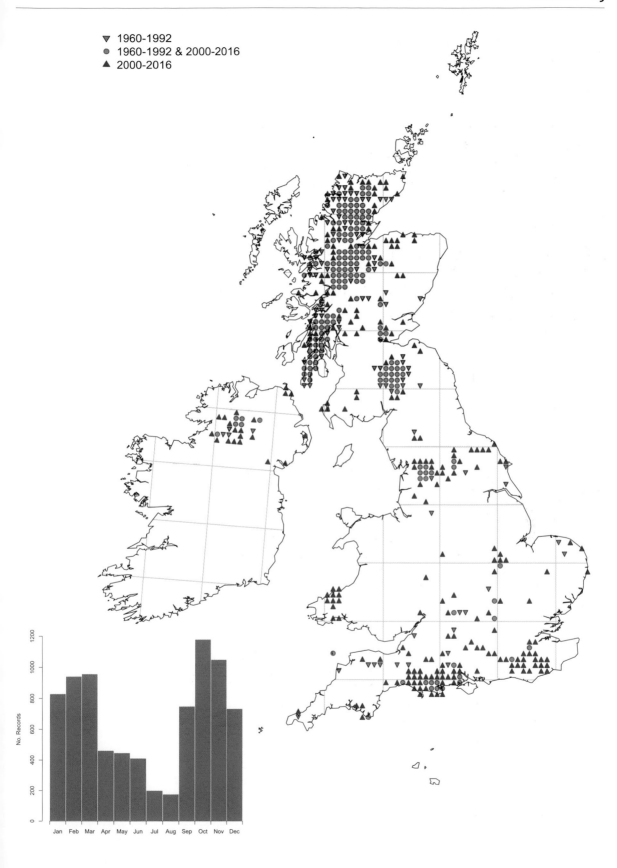

Fallow deer
Dama dama (LINNAEUS, 1758)

DES HAWLEY

DISTRIBUTION

The fallow deer is naturalised in Great Britain and Northern Ireland. The current population is largely descended from deliberate introductions and park escapes. Now, the species is widely distributed and locally abundant across much of England and Wales, while it is patchier and more localised in Scotland and Northern Ireland. Records are derived from the British Deer Society five-yearly Deer Distribution Survey (2016 data), together with other sightings, camera-trap images and road casualties.

ECOLOGY

The fallow deer is a grazing and browsing herbivore that prefers mature broadleaved woodland with understorey, open coniferous woodland and open agricultural land. It is largely diurnal but can become more nocturnal in disturbed areas. The fallow deer is a herding species, tending to spend most of the year in single-sex groups which can occasionally number as many as a hundred animals, though smaller groups are more usual. In larger populations, mature males (bucks) and females (does) usually come together only for the annual rut, which peaks in mid-October. A single fawn is born in June or July; twins are rare.

IDENTIFICATION

Similar in size to the sika deer, the fallow deer stands up to 100 cm high at the shoulder, though sexual dimorphism is apparent with the bucks being significantly larger. The fallow deer is the only deer in the UK with naturally occurring coat variations, ranging from black to white, some of which predominate according to location. The common variety is a chestnut colour with white spots in summer coat, changing to grey in winter, with a white rump bordered by a black inverted horseshoe marking. The black and white varieties are uniform in colour with no distinct rump markings. The tail is long and mobile. The buck grows large palmated antlers, which most animals shed during April. New sets of antlers are progressively larger with age.

BIBLIOGRAPHY

BRITISH DEER SOCIETY. 2017. 2016 deer distribution survey. Accessed at https://www.bds.org.uk/index.php/news-events/297-2016-deer-distribution-survey-results-now-available (30 July 2019).

PARLIAMENTARY OFFICE OF SCIENCE AND TECHNOLOGY. 2009. Wild deer. POSTnote No. 325. Accessed at http://researchbriefings.parliament.uk/ResearchBriefing/Summary/POST-PN-325 (30 July 2019).

AUTHORS   Jochen Langbein and Charles Smith-Jones

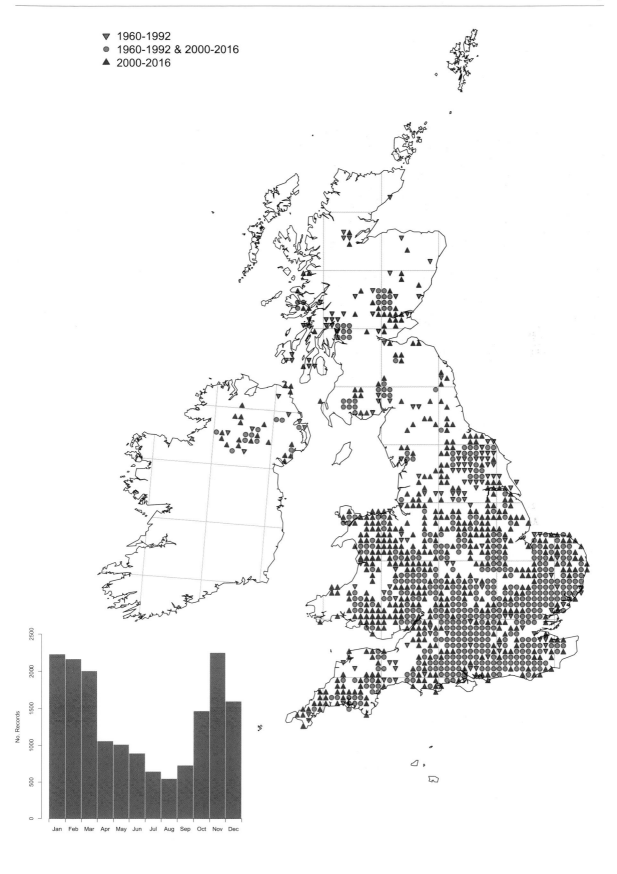

Chinese water deer
Hydropotes inermis (SWINHOE, 1870)

MARK BALDWIN

DISTRIBUTION

The Chinese water deer is native to eastern China, the Republic of Korea and the Democratic People's Republic of Korea, where populations are considered to be fragmented and decreasing (listed as Vulnerable on the International Union for the Conservation of Nature Red List). The UK population forms a high proportion of the global total. Deliberate and accidental introduction to locations in England occurred during the last century. The Chinese water deer is mainly found in Bedfordshire, Buckinghamshire and East Anglia, with some range expansion mainly to the west and south. There are small isolated populations elsewhere, probably attributable to escapes from collections and deliberate releases. There has been no confirmed record of wild Chinese water deer in Scotland, Wales or Northern Ireland. Records are derived from the British Deer Society five-yearly Deer Distribution Survey (2016 data), together with other sightings, camera-trap images and road casualties.

ECOLOGY

A selective browsing and grazing herbivore, the Chinese water deer prefers wetland habitats but also uses parks and farmland in the UK. It is usually seen alone or in small groups, and is mostly active from dusk until dawn. The rut takes place in late November and December,

and young are born in May and June. The Chinese water deer can give birth to up to six fawns, but generally produces two or three. Mortality can be very high during the first few weeks of life.

IDENTIFICATION

The Chinese water deer is a small deer, 50–55 cm high at the shoulder. It is slightly larger than the Reeves' muntjac deer but is smaller than the roe deer. The coat is a uniform russet brown in summer and is paler in winter. It has large, hairy and rounded ears, and the face is said to resemble a teddy bear. The Chinese water deer has no distinctive rump markings and has a very short stubby tail. The rump is held higher than the shoulders. This species never grows antlers; instead, mature males have long, visible canine tusks, which are used for fighting with other males. It can be difficult to distinguish between the sexes in the field.

BIBLIOGRAPHY

BRITISH DEER SOCIETY. 2017. 2016 deer distribution survey. Accessed at https://www.bds.org.uk/index.php/news-events/297-2016-deer-distribution-survey-results-now-available (30 July 2019).

HARRIS, R.B. & DUCKWORTH, J.W. 2015. *Hydropotes inermis*. *The IUCN Red List of Threatened Species* 2015: e.T10329A22163569. Accessed at http://dx.doi.org/10.2305/IUCN.UK.2015-2.RLTS.T10329A22163569.en (30 July 2019).

COOKE, A.S. 2012. Chinese puzzle. *Deer* 16 (2): 10–14.

AUTHORS   Arnold Cooke and Charles Smith-Jones

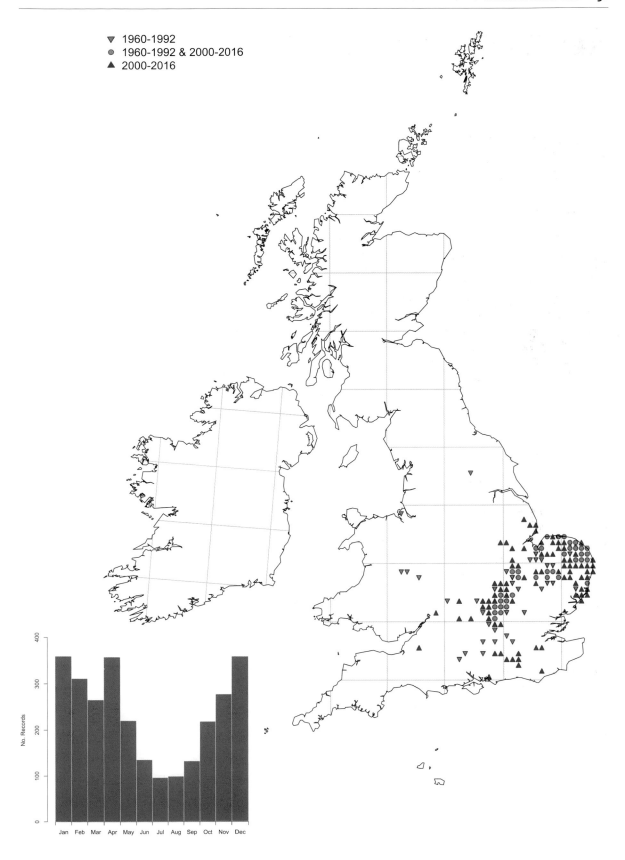

1960-1992
1960-1992 & 2000-2016
2000-2016

Greater horseshoe bat

Rhinolophus ferrumequinum (SCHREBER, 1774)

STEPHANIE COOLING-GREEN

DISTRIBUTION

Following population declines in the early and mid-twentieth century, by the 1960s the greater horse-shoe bat was restricted to south-west England and the southern coastal counties of Wales. It now shows signs of recovery with population numbers in the core of its range doubling in the past 20 years. In the same period, its distribution has expanded to include north and mid-Wales and the remaining southern counties of England. It has also spread up into southern areas of the west Midlands. Records in these areas are generally limited to individual animals, although a small number of summer colonies have also been discovered and there is evidence of some breeding. It is not recorded in Scotland or Northern Ireland.

ECOLOGY

During the summer, the greater horseshoe bat forms maternity colonies, with roosts that are generally located in the attics of old buildings, although occasionally it will also breed in underground sites. During winter, this bat will hibernate in caves, mines and other underground sites. It is usually associated with landscapes of decid-uous woodland with grazing pastures, in areas of the country where there are caves and mines. The species displays seasonal variation in its preferred foraging habitat. In the summer, cattle-grazed pasture is the favoured habitat where it feeds along field margins on emerging dung beetles. During the spring and autumn, it is more associated with semi-natural ancient wood-land. The greater horseshoe bat feeds preferentially on larger insect prey such as beetles or moths but will take smaller insect prey if the availability of the larger ones is limited. This species avoids flying in open or well-lit habitat, consequently commuting to its foraging grounds along linear landscape features such as well-grown hedgerows and avoiding areas with artificial lighting.

IDENTIFICATION

One of the largest British bat species with a body the size of a pear, and easily identified by its horseshoe-shaped nose-leaf. It has buff-brown fur often tinged red. In torpor or hibernation, the wings are loosely wrapped around the body.

BIBLIOGRAPHY

BRIGGS, P., HAWKINS, C., SHEPPARD, T. & WILSON, B. 2018. *The State of the UK's Bats 2017 (National Bat Monitoring Programme Population Trends)*. London: Bat Conservation Trust.
DIETZ, C. & KIEFER, A. 2016. *Bats of Britain and Europe*. London: Bloomsbury.

AUTHOR   Henry Schofield

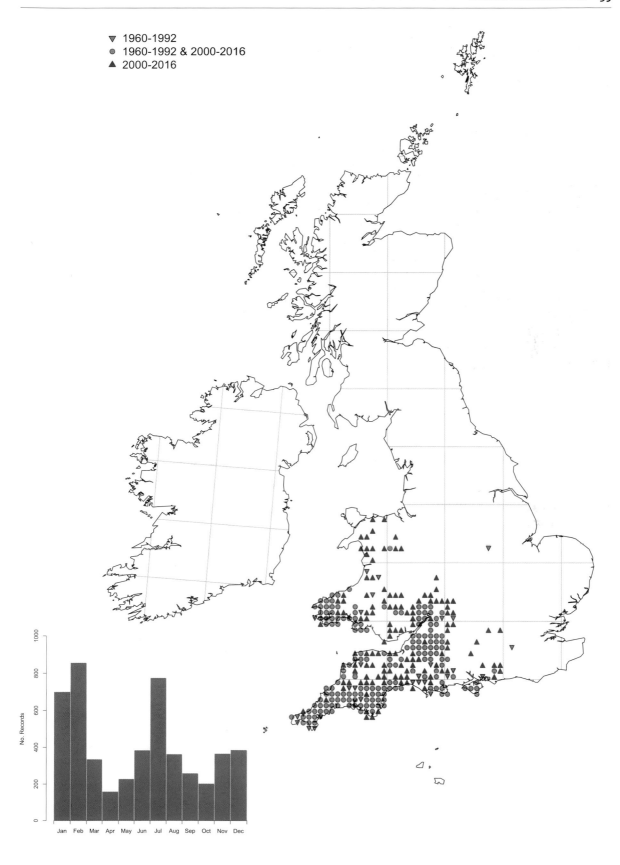

Alcathoe bat

Myotis alcathoe (VON HELVERSEN & HELLER, 2001)

CHRISTOPHER DAMANT

DISTRIBUTION

The Alcathoe bat was only confirmed in the UK in 2010 as it was previously not distinguished from whiskered *Myotis mystacinus* or Brandt's *Myotis brandtii* bats. Therefore no historical datasets or distribution records exist. This highly cryptic species is difficult to identify and survey effort is very low, so it is likely to be under-recorded and its true distribution is likely to be larger than that currently known. Nevertheless, like populations elsewhere in Europe, its distribution appears to be patchy.

Good populations are present throughout Sussex and Surrey where a number of maternity roosts have been identified. Breeding females have also been caught in Yorkshire, west Kent and Jersey. Individuals have most recently been discovered in Hampshire (2017) and at an autumn swarming site in Wiltshire (2018). It has not been recorded in Wales, Scotland or Northern Ireland.

Most records are from trapping surveys, and commonly require DNA analysis to confirm species identification of animals judged likely to be Alcathoe bat on the basis of morphometric features.

ECOLOGY

The Alcathoe bat appears to be a woodland specialist. It is mainly found foraging in broadleaved woodland, notably ancient and mature oak woodland, but also in woodland edges and wet woodlands, and, to a lesser extent, scrub and parkland habitats.

The maternity roosts identified to date are almost exclusively located in the splits, cracks and loose bark of trees within woodlands, but one building roost is known in Surrey. It has not been recorded using bat boxes. Colonies appear to fragment into a number of small satellite roosts, and it is common for only low numbers to roost together at any one time (peak colony count of 96 individuals). Although it is captured at swarming sites, there is only one record of this species hibernating underground.

IDENTIFICATION

The Alcathoe bat is the smallest of the *Myotis* bat species, with the whiskered and Brandt's bats being very marginally larger. It has the typical pattern of fur colouration for *Myotis* bats, having pale underparts, darker fur on the upper surfaces and a pointed tragus. The Alcathoe bat can be distinguished from the whiskered or Brandt's bat by its paler face, ears and tragus, its blunter muzzle and its shorter tragus to ear notch height.

BIBLIOGRAPHY

WHITBY, D. *pers. obs.*

AUTHOR   Daniel Whitby

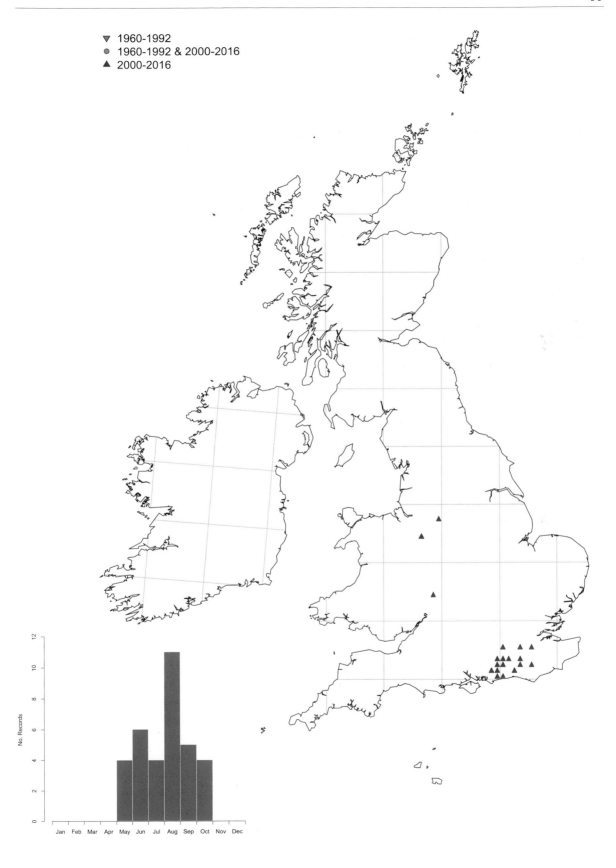

- ▼ 1960-1992
- ● 1960-1992 & 2000-2016
- ▲ 2000-2016

Brandt's bat

Myotis brandtii (EVERSMANN, 1845)

PAUL KENNEDY

DISTRIBUTION

As a cryptic small *Myotis* species, the Brandt's bat was only recognised as a species separate from the whiskered bat in 1970 in the UK. It is considered widespread throughout England, although is largely absent from East Anglia and patches along the north-east coast. It is more common in northern England and becomes rarer moving south, with no confirmed maternity colonies in many southern counties. The Brandt's bat is present throughout Wales and has very recently been identified in southern Scotland. It appears to be absent elsewhere in Scotland and in Northern Ireland.

The Brandt's bat cannot be distinguished from the whiskered bat using acoustic analysis, and most confirmed species records are from trapping surveys, or from DNA analysis of droppings.

ECOLOGY

The echolocation and morphological characteristics of the Brandt's bat are very similar to those of the whiskered and Alcathoe bats. It is highly manoeuvreable in flight and has a broad dietary range, feeding as a hawker and gleaner of flies, moths and spiders.

The species uses a wide range of habitats including coniferous, mixed and broadleaved woodland as well as grassland and scrub, but it generally avoids urban areas. This species is frequently caught commuting along linear features and has been recorded foraging up to 3.2 km from the roost.

Most identified maternity roosts are in buildings, but they also occur in bat boxes, bridges and trees. Under-recording of tree roosts is highly likely. It uses swarming sites in late August and early September, and is commonly found hibernating underground.

IDENTIFICATION

The largest of the three small *Myotis* species, the Brandt's bat has similar acoustic and morphological characteristics to the whiskered and Alcathoe bats. There are no morphological features that categorically separate the small *Myotis* species, even in the hand, and DNA analysis is regularly used to identify these species. However, the Brandt's bat commonly has a large bulbous penis and a high protocone in the dentition (fourth premolar, upper jaw) which are the most reliable and commonly used identification features.

BIBLIOGRAPHY

BERGE, L. 2007. Resource partitioning between the cryptic species Brandt's bat (*Myotis brandtii*) and the whiskered bat (*M. mystacinus*) in the UK. Unpublished PhD thesis, University of Bristol.
VAUGHAN, N. 1997. The diets of British bats (*Chiroptera*). *Mammal Review* 27 (2): 77–94.
WHITBY, D. *pers. obs.*

AUTHOR Daniel Whitby

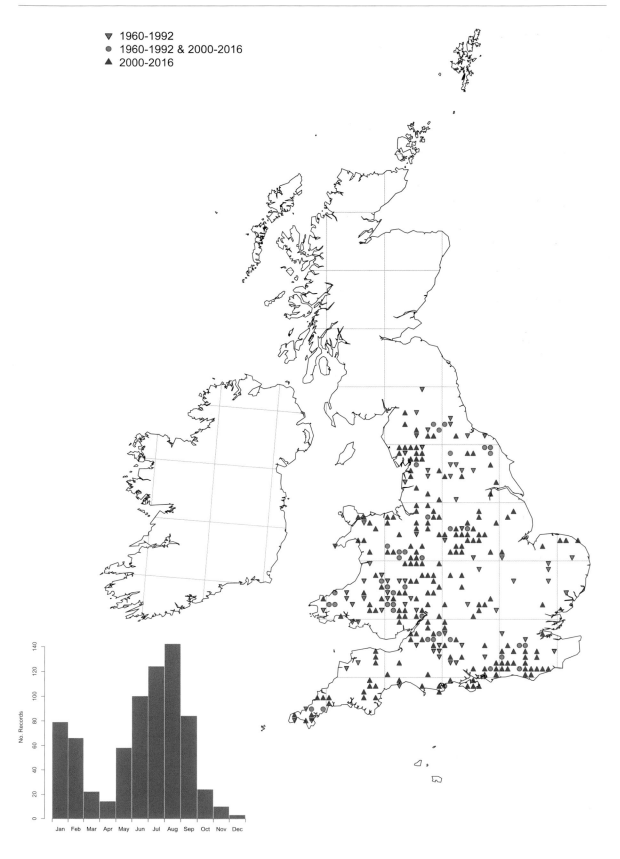

Daubenton's bat

Myotis daubentonii (KUHL, 1817)

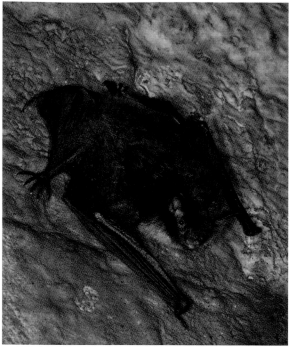

HENRY SCHOFIELD

DISTRIBUTION

The Daubenton's bat is widespread across the UK. Its distribution appears patchier in Scotland and Northern Ireland, although it is not clear whether this represents true absence or is due to lower recording effort. Records are derived from a combination of visual surveys of bats in flight (identified by their characteristic hunting behaviour over water surfaces), roost visits, acoustic surveys, grounded animals and the capture of individuals in the field. There is no evidence of a change in distribution over time.

ECOLOGY

The Daubenton's bat usually forages low over smooth water, taking insects (mainly aquatic flies) from the air, but also 'gaffing' prey from the water's surface using its large feet or tail membrane. Identified summer roosts are typically close to water, in bankside trees, buildings and bridges, but the species is highly under-recorded. Roost switching is common during the breeding season,

and the sexes are usually segregated. The species is frequently recorded swarming at underground sites in the autumn (August), a behaviour likely to be associated with mating. The species is usually found hibernating in caves and other underground sites.

IDENTIFICATION

The Daubenton's bat is medium-sized relative to other UK bats (6–10 g). It has short ears for a *Myotis* species and a relatively blunt tragus and muzzle. Its dorsal fur is brown, and it has whitish grey fur on its ventral side. The calcar is straight and extends more than two-thirds along the length of the tail membrane. It also has large feet, which are covered with stiff bristles. As with other *Myotis* bats, it can be difficult to identify the species conclusively on the basis of acoustic records alone.

BIBLIOGRAPHY

ANGELL, R.L., BUTLIN, R.K. & ALTRINGHAM, J.D. 2013. Sexual segregation and flexible mating patterns in temperate bats. *PLoS ONE* 8 (1): e54194.
SENIOR, P., BUTLIN, R.K. & ALTRINGHAM, J.D. 2005. Sex and segregation in temperate bats. *Proceedings of the Royal Society B* 272 (1580): 2467–2473.

AUTHOR   Anita Glover

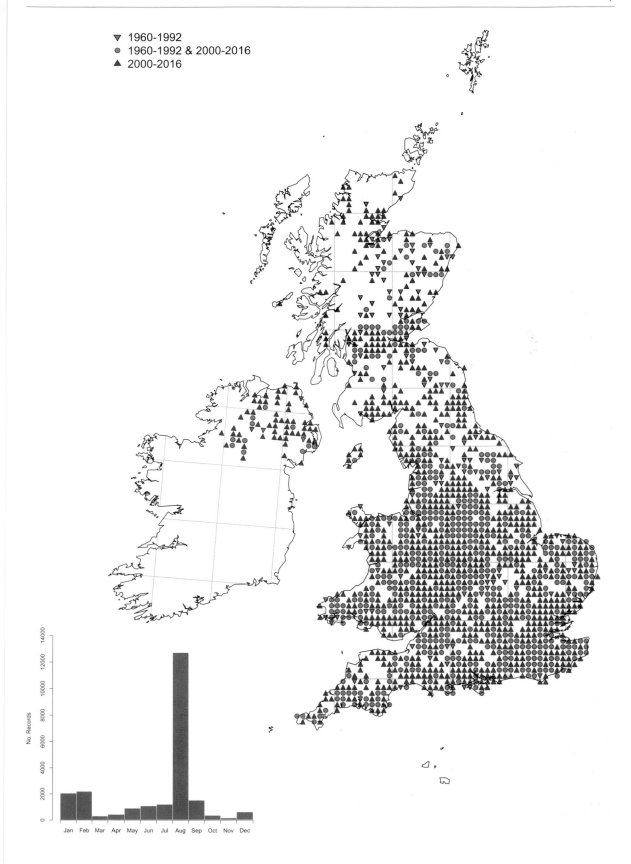

▼ 1960-1992
● 1960-1992 & 2000-2016
▲ 2000-2016

Natterer's bat
Myotis nattereri (KUHL, 1817)

CHRIS DAMANT

ECOLOGY

The Natterer's bat is commonly associated with trees, particularly broadleaved woodlands, tree-lined river corridors, trees in parkland, and hedgerows adjacent to pasture. It has also been observed foraging over grass and thistles on roadsides, in the open over pasture and meadows, and using mature Corsican pine plantations in Scotland.

Maternity roosts are located in trees, bat boxes and buildings – largely barns, churches and old dwelling houses. Roosts tend to be close to woodlands, with most being within 500 m; however, the size of the woodland does not appear important. Winter roosts may be found in underground sites such as canal and railway tunnels, caves, mines and ice houses.

IDENTIFICATION

The Natterer's bat is a medium-sized *Myotis* bat with brown-grey dorsal fur and grey-white under and ventral fur. The main distinguishing characteristics are the S-shaped calcar, the conspicuous fringe of short stiff hairs along the edge of the tail membrane, and the long, sharply pointed tragus. The ears are relatively long, reaching a little beyond the tip of the muzzle if folded forwards. As with other *Myotis* bats, it can be difficult to identify the species conclusively on the basis of acoustic records alone.

BIBLIOGRAPHY

SMITH, P.G. & RIVERS, N.M. 2008. Natterer's bat *Myotis nattereri*. In S. Harris & D.W. Yalden (eds) *Mammals of the British Isles*, 4th edn. Southampton: The Mammal Society. pp. 323–328.
STEBBINGS, R.E. 1993. *The Greywell Tunnel: An Internationally Important Haven for Bats.* Newbury: English Nature.

AUTHOR   Peter Smith

DISTRIBUTION

The Natterer's bat is widespread in the UK. Its distribution is sparser in Scotland and Northern Ireland than in England and Wales, but some under-recording is likely. It is also found on the island of Arran, the Isle of Man and the Isle of Wight. There is a single recent record from Islay. Records are derived from roost visits, grounded bats and trapping surveys.

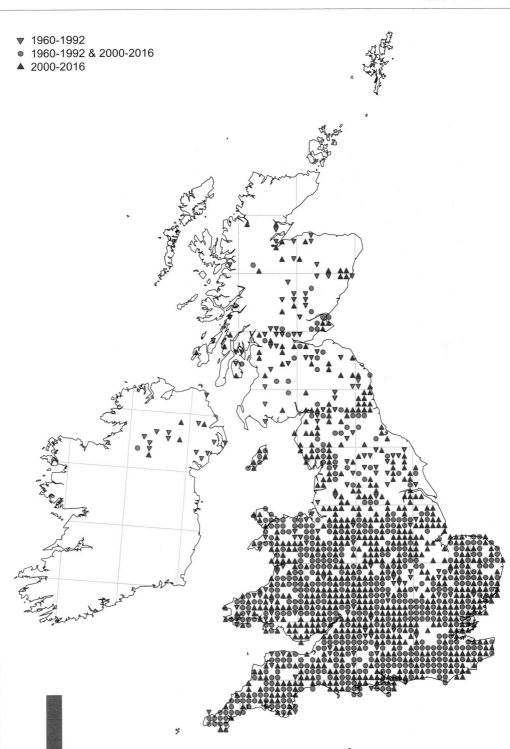

Serotine bat

Eptesicus serotinus (SCHREBER, 1774)

DANIEL HARGREAVES

DISTRIBUTION

The serotine bat occurs across southern England, with scattered records in Wales. It is absent from Scotland and Northern Ireland. A high proportion of the total records are derived from acoustic surveys, and the increasing use of broadband acoustic detectors over time is likely to explain the apparent increase in occupancy of this species. Because of overlap in call parameters, acoustic records from noctule *Nyctalus noctula* and Leisler's *Nyctalus leisleri* bats may be confused for those of the serotine bat. Other records come from roost visits, trapping surveys and grounded animals. Expert opinion suggests a decline in roost occupation in south-east England with a corresponding increase in the south-west.

ECOLOGY

The serotine bat is associated with pastures and parklands. It is an early emerging species and feeds on moths, beetles and flies. Summer roosts are usually in the roofs and walls of large houses, and maternity colonies tend to be small. The animal is quiet and so roosts in buildings are likely to be under-recorded. Few winter roosts are known, though they are occasionally found at underground sites where they are also sometimes captured during swarming.

IDENTIFICATION

One of the larger British bats, the serotine bat has a rounded tragus and a dark muzzle, ears and wing membrane. A post-calcarial lobe is present and there is a free tail tip. The fur is dark and long compared with that of the noctule bat – the only likely confusion in the hand. Droppings are large and may be confused with those of greater horseshoe bats, but are oval in profile.

BIBLIOGRAPHY

HUTSON, A.M. 2008. Serotine *Eptesicus serotinus*. In S. Harris & D.W. Yalden (eds) *Mammals of the British Isles*, 4th edn. Southampton: The Mammal Society. pp. 356–360.

AUTHOR   Fiona Mathews

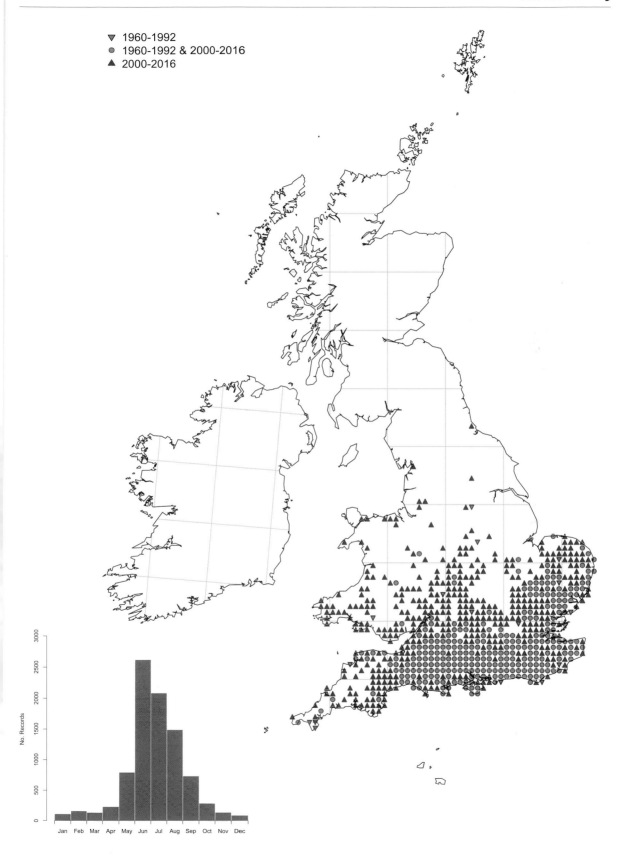

▼ 1960-1992
● 1960-1992 & 2000-2016
▲ 2000-2016

Common pipistrelle bat
Pipistrellus pipistrellus (SCHREBER, 1774)

PAUL KENNEDY

DISTRIBUTION

The common pipistrelle bat is present across the UK. Gaps in the distribution will usually reflect a lack of survey effort, since this bat is found across all habitats even at high altitudes (e.g. upland wind energy sites) and in highly urban environments. Records are derived from acoustic surveys, roost visits, grounded animals and capture. Time trends in distribution are difficult to assess because the species was only recently distinguished from the soprano pipistrelle bat *Pipistrellus pygmaeus*. Although all records used here are for *P. pipistrellus sensu stricto*, and those submitted with a generic term (e.g. pipistrelle bat) are excluded, older records will be a combination of common and soprano pipistrelle bats. Even now, misidentification is likely because of similarities in morphology, habitat and geographical ranges. The increasing use of acoustic detectors, coupled with extensive survey effort for built developments over the last 15 years, is likely to explain the apparent range expansion between the historical and current atlas periods.

ECOLOGY

The species is found in all habitat types. It emerges from the roost early and feeds mainly on flies. Summer roosts are mainly in buildings, including barns, churches and domestic houses. Maternity colonies are large, sometimes including several hundred animals, and are noisy. Roost switching within a season is common. Few winter roosts are known, though they are very rarely found at underground sites and in buildings.

IDENTIFICATION

The common pipistrelle bat is one of the smallest British bats, weighing just 4–6 g. Like other pipistrelle species, it has a rounded tragus and short, blunt ears. The fur is mid-brown, with little differentiation between the dorsal and ventral side. A suite of characteristics including smell, nostril shape, wing veination and call profile can aid separation from the soprano pipistrelle bat. Droppings are small, and may potentially be confused with those of small *Myotis* bats. It can be distinguished acoustically from other pipistrelle bats reasonably reliably on the basis of the peak frequency of its call.

BIBLIOGRAPHY

BARRATT, E.M., DEAVILLE, R., BURLAND, T.M., BRUFORD, M.W., JONES, G., RACEY, P.A. & WAYNE, R.K. 1997. DNA answers the call of pipistrelle bat species. *Nature* 387 (6629): 138–139.
LINTOTT, P.R., BARLOW, K., BUNNEFELD, N., BRIGGS, P., GAJAS ROIG, C. & PARK, K.J. 2016. Differential responses of cryptic bat species to the urban landscape. *Ecology and Evolution* 6 (7): 2044–2052.
MATHEWS, F., RICHARDSON, S.M. & HOSKEN, D.J. 2016. *Understand the Risks to Bat Populations Posed by Wind Turbines – Phase 2 – WC0753*. London: DEFRA.

AUTHOR   Fiona Mathews

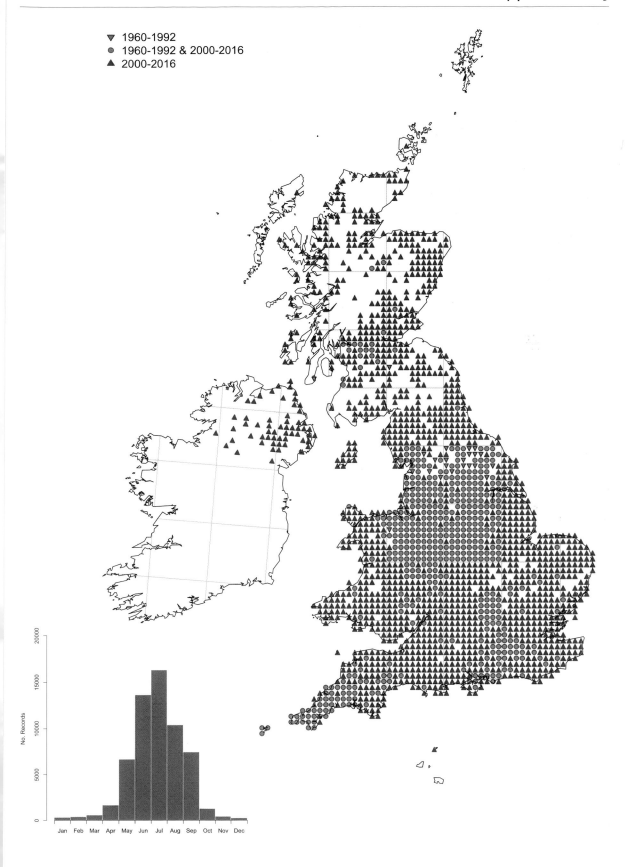

Barbastelle bat
Barbastella barbastellus (SCHREBER, 1774)

CHRISTOPHER DAMANT

DISTRIBUTION

The barbastelle bat is found throughout the southern half of the UK and is absent from northern England, Scotland and Northern Ireland. Records reach up to north Wales and across to Lincolnshire, with occasional older records stretching up to North Yorkshire. There are good populations throughout much of East Anglia, Sussex, Hampshire and the Isle of Wight, the south Wales border counties and much of south-west England. However, it has a low population density and it is infrequently encountered in trapping surveys. It is likely to be under-recorded throughout its range as it rarely roosts in buildings and so detection relies on specialist acoustic and trapping surveys.

ECOLOGY

Maternity roosts are almost exclusively found in trees (cracks, splits, hazard beams, wounds and lose bark), particularly oaks in ancient woodland but also in individual parkland trees. The species only rarely has maternity roosts in buildings in the UK, though solitary individuals are sometimes found. The barbastelle bat usually hibernates in trees, particularly in areas with a dense understorey, so these roosts are rarely found. However, occasional individuals are recorded in underground sites including caves, tunnels and grottos. It is caught in low numbers at swarming sites, although it is rarely found hibernating at these same locations.

The barbastelle bat forages over wide areas and can use a range of habitats. During the breeding season it forages in woodlands and close to large hedgerows early in the night, but later in the night it can use floodplains, meadows and unimproved grasslands, and readily crosses open areas. Radio tracking in southern England has shown that the mean core range of females from maternity colonies is 8 km, but they can fly long distances rapidly, frequently crossing very open habitat including downland and moorland, to reach foraging areas up to 20 km away. This species is an aerial hawker that feeds almost exclusively on hearing moths, and it is likely that its extensive foraging range is linked with food availability. During cooler periods, it is commonly found foraging in riparian and old broadleaved woodland, notably oak with a good understorey.

IDENTIFICATION

The barbastelle bat is a medium-sized bat impossible to confuse with other European species, so is easily identified in the hand. It has darkly coloured skin and a short 'pug'-shaped nose, with the nostrils opening upwards. The fur is dark, often with light frosted tops on its back and a paler ventral side. The eyes appear to be set inside the base of the ears. The ears are triangular and meet in the middle of the head, a feature shared only with the long-eared bats *Plecotus* spp.

BIBLIOGRAPHY

GREENAWAY, F. 2001. The Barbastelle in Britain. *British Wildlife* 12 (5): 327–334.

ZEALE, M.R.K. 2011. Conservation biology of the barbastelle (*Barbastella barbastellus*): applications of spatial modelling, ecology and molecular analysis of diet. Unpublished PhD thesis, University of Bristol.

ZEALE, M.R.K., DAVIDSON-WATTS, I. & JONES, G. 2012. Home range use and habitat selection by barbastelle bats (*Barbastella barbastellus*): implications for conservation. *Journal of Mammalogy* 93 (4): 1110–1118.

AUTHOR   Daniel Whitby

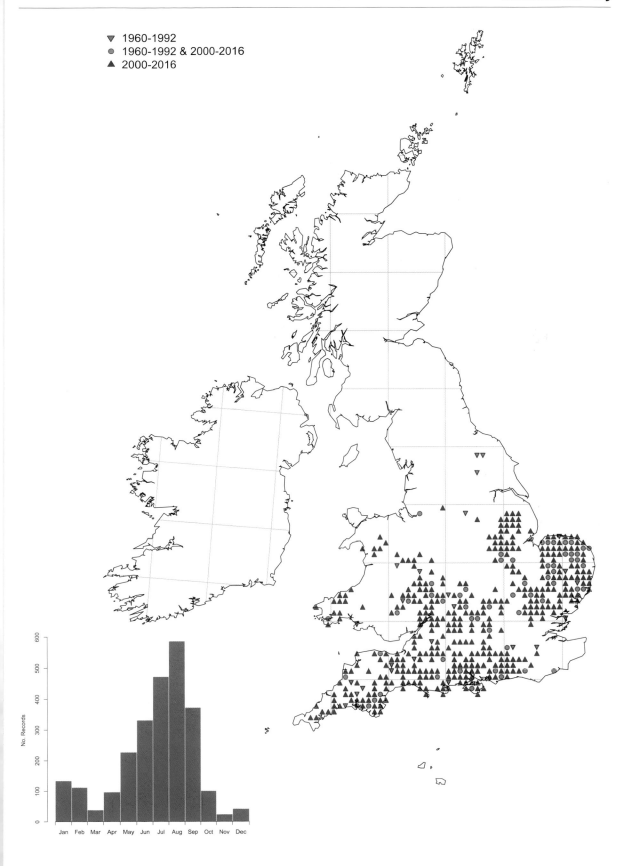

▽ 1960-1992
● 1960-1992 & 2000-2016
▲ 2000-2016

Harbour seal

Phoca vitulina (LINNAEUS, 1758)

DEREK CRAWLEY

DISTRIBUTION

Approximately 30% of the harbour seal population of Europe is found in the UK, but this proportion has declined from approximately 40% in 2002. The harbour seal is widespread around the west coast of Scotland and throughout the Hebrides and Northern Isles. It is regularly found in Northern Ireland and the largest breeding population is in Strangford Lough. On the east coast of the UK, its distribution is more restricted, with concentrations in the major estuaries of the Thames, the Wash, the Firth of Tay and the Moray Firth. Approximately 79% of the UK harbour seal population is in Scotland, with 16% in England and 5% in Northern Ireland. Major declines have now been documented in several populations around Scotland since 2000. The causes of the recorded declines are unknown, but are not thought to be linked to the 2002 phocine distemper virus epidemic. However, the Wash and eastern England were not severely affected by this disease until 2009.

ECOLOGY

The harbour seal shows little sexual dimorphism with adults typically weighing 80–100 kg. It is a long-lived animal, and reaches 20–30 years of age.

The species comes ashore in sheltered waters, typically on sandbanks and in estuaries, but also in rocky areas. Most haul-out sites are used daily, although foraging trips can last for several days. The harbour seal is primarily a coastal species and normally feeds within 40–50 km of its haul-out sites. It is a generalist feeder that takes a wide variety of fish, cephalopods and crustaceans that are obtained from surface, mid-water and benthic habitats. Its diet varies seasonally and geographically but includes sandeels, gadoids, herring and sprat, flatfish, octopus and squid.

It gives birth to pups in June and July. The pups are born having shed their white coat in utero and can swim almost immediately. Females come into oestrus about a month after giving birth, and mating takes place in the water with males defending underwater calling territories. The moult in August follows the pupping and mating season, and the timing of the moult varies by age and sex class.

IDENTIFICATION

The harbour seal has a small, catlike head with V-shaped nostrils. The eyes are relatively large and set close together. Prominent, light-coloured vibrissae are characteristic of the species, and it has a fine, spot-patterned pelage that can vary in colour from grey to brown, with some ring-like markings and larger blotches.

BIBLIOGRAPHY

BONESS, D.J., BOWEN, W.D., BUHLEIER, B.M. & MARSHALL, G.J. 2006. Mating tactics and mating system of an aquatic-mating pinniped: the harbor seal, *Phoca vitulina*. *Behavioural Ecology and Sociobiology* 61 (1): 119–130.

SCOS, 2016. *Scientific Advice on Matters Related to the Management of Seal Populations: 2016.* SCOS Main Advice 2016. Swindon: Scientific Committee on Seals.

TOLLIT, D.J., BLACK, A.D., THOMPSON, P.M., MACKAY, A., CORPE, H.M., WILSON, B., VAN PARIJS, S.M., GRELLIER, K. & PARLANE, S. 1998. Variations in harbour seal *Phoca vitulina* diet and dive-depths in relation to foraging habitat. *Journal of Zoology* 244 (2): 209–222.

AUTHOR   David Thompson

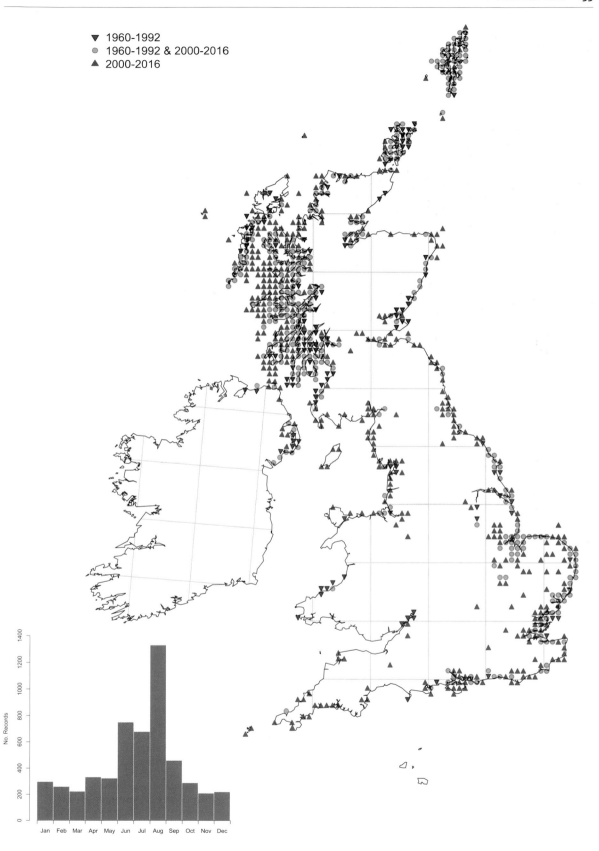

- ▼ 1960-1992
- ● 1960-1992 & 2000-2016
- ▲ 2000-2016

No. Records

Bowhead whale

Balaena mysticetus (LINNAEUS, 1758)

CAROLINE WEIR

DISTRIBUTION

The bowhead whale is an Arctic species with a circum-polar distribution. It favours the ice edge around the Arctic Ocean and the waters surrounding Greenland and Svalbard across to Novaya Zemlya, migrating northwards in summer into the high Arctic as the ice retreats. In recent years, however, the species has been recorded much further south – in Cape Cod Bay, New England (eastern USA), as well as off the coast of Cornwall (Isles of Scilly in February 2015, Marazion in May 2016), Northern Ireland (County Down in May 2016), France (Brittany in May 2016), Belgium (Ostende in March–April 2017), the Netherlands (Vlissingen in April 2017) and Ireland (County Cork in April 2017). The European sightings could all represent the same lone individual. These extralimital records may reflect the break-up and southward drift of ice that the Arctic is experiencing.

ECOLOGY

The bowhead whale occupies Arctic and subarctic regions of the northern hemisphere, generally between 55° and 85°N.

IDENTIFICATION

The bowhead whale only superficially resembles its relative, the North Atlantic right whale. It has a very rotund body, 18–20 m in length, a large (up to *c.* 40% of its body) though relatively narrow head and a strongly arched lower jaw. Like other right whales, it is black in colour and has no dorsal fin. A distinctive feature is the prominent muscular bulge in the area of the blowhole with an obvious depression behind. It has a very thick skin and blubber layer, and the longest baleen plates of any whale. The plates are dark grey to brown or black, generally with lighter fringes, which from the front may show as a white patch in front of the lower jaw. The flippers are large and fan-shaped with blunt tips. The tail flukes are wide and tapered at the tips with no central notch, and there is often a light grey or white band across the tailstock.

BIBLIOGRAPHY

GEORGE, J.C., RUGH, D. & SUYDAM, R. 2017. Bowhead whale *Balaena mysticetus*. In B. Würsig, J.G.M. Thewissen & K.M. Kovacs (eds) *Encyclopedia of Marine Mammals*, 3rd edn. London: Academic Press. pp. 133–135.

AUTHORS   Peter Evans

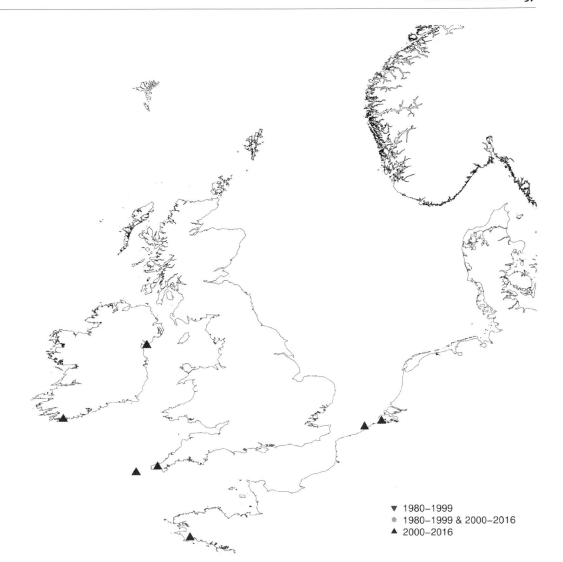

▼ 1980–1999
● 1980–1999 & 2000–2016
▲ 2000–2016

Blue whale

Balaenoptera musculus (LINNAEUS, 1758)

PETER EVANS, SEA WATCH FOUNDATION

DISTRIBUTION

The blue whale is found worldwide in all seas. Although rare after a century of exploitation, the species occurs regularly in deep waters of the North Atlantic, from the Caribbean to the Davis Strait/southern Greenland in the west, and from the Canaries, Cape Verde islands and West Africa to Jan Mayen, Svalbard and the Barents Sea in the east. There have been no sightings in the shelf seas of Britain and Ireland. All records are from deep waters offshore between May and October. Sightings and acoustic monitoring both reveal small numbers of animals in the deep waters of the Faroe–Shetland Channel and Rockall Trough, south to the Bay of Biscay.

ECOLOGY

The blue whale is usually found in deep waters of 400–1,000 m in depth. It is thought to spend the winter in tropical and subtropical seas where it breeds, and then migrates to feed during summer months in cold temperate and polar waters. However, acoustic recordings in the mid-Atlantic suggest that some individuals, at least, remain at high latitudes throughout winter.

IDENTIFICATION

At 23–27 m in length, the blue whale is the largest of all mammals. The head is broad, flat and U-shaped, with a single ridge extending from a raised area forward of the blowhole towards the tip of the snout. The head and most of the body is characteristically pale bluish-grey, mottled with grey or greyish white. It has a very small dorsal fin that varies in shape from nearly triangular to moderately recurved, and is situated distinctly more than two-thirds along the back so that it is seen only just prior to a dive, and sometimes after the blow. The blow itself is tall and slender, rising vertically to a height of 9–12 m. On diving, the species lifts its tail at a slight angle, whereas in the fin *Balaenoptera physalus*, sei *Balaenoptera borealis* and minke *Balaenoptera acutorostrata* whales, the tail is rarely visible above the surface. The tail is broad and triangular in shape, with slender, pointed tips to the flukes and only a slight central notch.

BIBLIOGRAPHY

EVANS, P.G.H. 2008. Blue whale *Balaenoptera musculus*. In S. Harris & D.W. Yalden (eds) *Mammals of the British Isles*, 4th edn. Southampton: The Mammal Society. pp. 675–678.

AUTHORS   Peter Evans

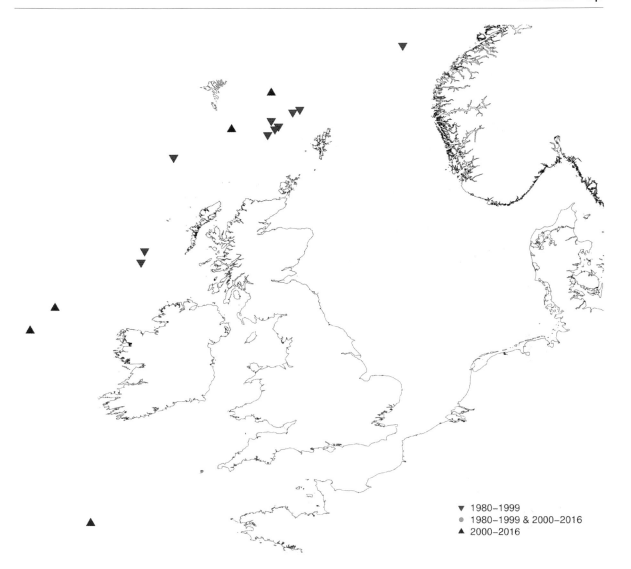

▼ 1980–1999
● 1980–1999 & 2000–2016
▲ 2000–2016

Sei whale

Balaenoptera borealis (LESSON, 1828)

CAROLINE WEIR

DISTRIBUTION

The sei whale has a worldwide distribution, occurring mainly offshore in deep waters from the tropics to polar seas of both hemispheres, with seasonal latitudinal migrations. In the central North Atlantic, summering populations are concentrated in deep waters north to Iceland, with a concentration of animals observed in June just north and south-west of the Charlie Gibbs Fracture Zone (53°N) over the Mid-Atlantic Ridge. It is also seen regularly in small numbers in the Azores and Madeira. In the western North Atlantic, the species is reported in summer mainly over the Nova Scotia shelf and in Labrador, and in winter from Florida, the Gulf of Mexico and the Caribbean. In the eastern North Atlantic, it is thought to winter off north-west Africa, Spain and Portugal and in the Bay of Biscay, migrating north to summering grounds off Shetland, the Faroes, Norway and Svalbard.

The sei whale is uncommon in UK and Irish waters, occurring mainly beyond the shelf edge between the Faroes and Northern Isles of Scotland, in the Rockall Trough south towards the Porcupine Bight. The species occurs occasionally in the shelf seas in the Hebrides, off Shetland and Orkney, north-east Britain, the south-west of England and western Ireland. All sighting records have been between June and October.

ECOLOGY

Appears to occur mainly in depths of 500–3,000 m.

IDENTIFICATION

The sei whale is a large slender rorqual typically 12–17 m in length with a relatively slender head. It has a slightly arched forehead, similar to that of the fin whale, but rounder than that of the blue whale. A single prominent ridge occurs along the middle of the top of the head, and the tip of the upper jaw tends to be down-turned. The head, back and flanks are usually a dark steely-black or dark grey colour. The blow of the sei whale resembles that of the fin whale but is lower and less dense, typically rising to a height of 3 m. The dorsal fin and blow tend to show almost simultaneously, and both remain in view for relatively long periods before a normally shallow dive. The dorsal fin is nearly erect and strongly recurved. It is taller than those of other large rorquals, and is placed slightly less than two-thirds along the back and thus further forward than that of the blue or fin whale.

BIBLIOGRAPHY

EVANS, P.G.H. 2008. Sei whale *Balaenoptera borealis*. In S. Harris & D.W. Yalden (eds) *Mammals of the British Isles*, 4th edn. Southampton: The Mammal Society. pp. 672–675.

AUTHORS Peter Evans

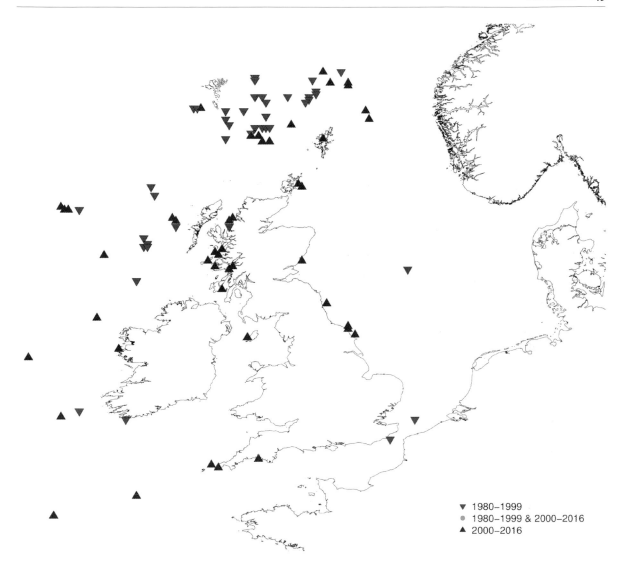

▼ 1980–1999
● 1980–1999 & 2000–2016
▲ 2000–2016

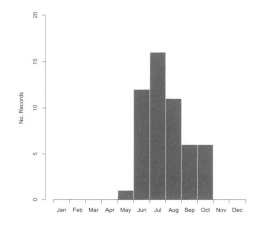

Northern bottlenose whale

Hyperoodon ampullatus (FORSTER, 1770)

SAANA ISOJUNNO

DISTRIBUTION

The northern bottlenose whale is confined to the North Atlantic from temperate to Arctic seas. Its main range extends from Baffin Island and west Greenland south to New England in the west, and from Svalbard to the southern tip of the Iberian Peninsula in the east, including around the oceanic archipelago of the Azores. It occurs casually further south (to *c.* 15°N), having been recorded in the Caribbean in the west and the Canaries in the east. Main regions of concentration, identified from former whaling activities, appear to be west of Norway, west of Svalbard, north of Iceland, in the Davis Strait off Labrador, off the Faroes, and in The Gully off eastern Canada. In north-west Europe, the species is seen primarily in waters exceeding 1,000 m, such as the Faroe–Shetland Channel, Rockall Trough and southern Bay of Biscay. In summer, it may move onto north-west European shelf waters, such as around the Hebrides, where most records occur between July and September.

ECOLOGY

The northern bottlenose whale typically inhabits deep ocean abysses of depths greater than 500 m, with greatest numbers in cold temperate to Arctic seas.

IDENTIFICATION

Being a deep diver (regularly to depths greater than 800 m), and able to remain under the surface for more than 70 minutes, the northern bottlenose whale is infrequently observed. Its most distinctive feature is the bulbous forehead and short dolphin-like beak that can be seen if the animal spy-hops or breaches as it occasionally does. The species frequently approaches vessels and has also been observed to lob-tail. The body is long (7–8.5 m in females; 8–9.5 m in males), robust and cylindrical in shape. Colouration varies from chocolate-brown to greenish-brown above, often lighter on the flanks, lightening to buff or cream all over with age. The blow is bushy, rising to a height of about 2 m, and is slightly forward pointing. The species has a strongly recurved dorsal fin two-thirds along the back, which may lead to confusion with the minke or sei whales unless other more distinctive features such as head shape are seen. Its fin is generally more erect and hooked than that of the minke whale, whereas an adult sei whale is almost twice the length of a bottlenose whale.

BIBLIOGRAPHY

HOOKER, S.K., GOWANS, S. & EVANS, P.G.H. 2008. Northern bottlenose whale *Hyperoodon ampullatus*. In S. Harris & D.W. Yalden (eds) *Mammals of the British Isles*, 4th edn. Southampton: The Mammal Society. pp. 685–690.

AUTHORS   Peter Evans

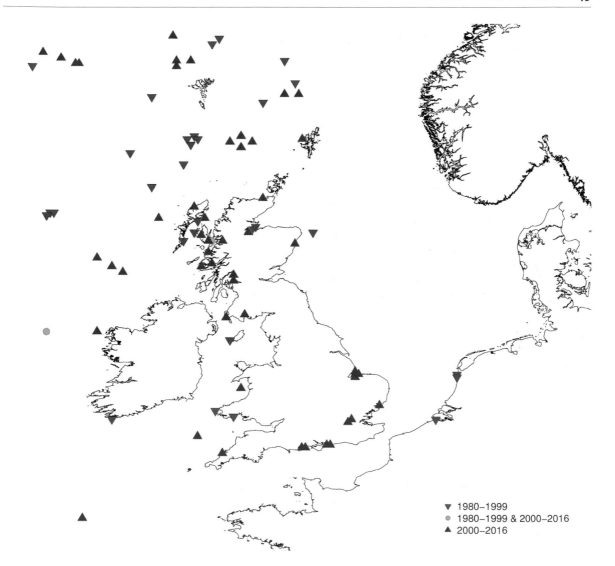

▼ 1980–1999
● 1980–1999 & 2000–2016
▲ 2000–2016

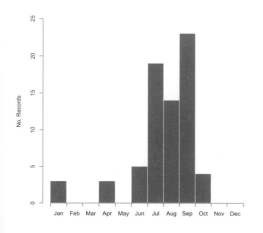

Cuvier's beaked whale

Ziphius cavirostris (CUVIER, 1823)

FRAZER COOMBER, CIMA RESEARCH FOUNDATION

DISTRIBUTION

The Cuvier's beaked whale has the most widespread distribution of the family Ziphiidae, probably occurring worldwide in tropical to warm temperate seas. Favouring deep, warm waters, Cuvier's beaked whale is the most common beaked whale in southern Europe – off the Iberian Peninsula, in the Bay of Biscay and in the Mediterranean. It is relatively rare in the UK and Ireland (although the number of records has increased in recent years) and further north and east, with isolated records from Iceland, Sweden and the Netherlands. It is a regular inhabitant of the waters around the Azores, Madeira and the Canaries, where it is seen all year round. In the western North Atlantic, it occurs in the Caribbean, Gulf of Mexico and south-east USA.

Around Britain and Ireland, there have been only a handful of sighting records in contrast to the number of sightings further south in the Bay of Biscay. On the other hand, there have been around 100 strandings on the coasts of Britain and Ireland over the last 100 years. Two-thirds of these strandings have occurred since 1963, almost all coming from the Atlantic coasts, mainly in late winter or spring. The few sightings in UK and Irish waters have been mainly in July, with none between September and December.

ECOLOGY

The Cuvier's beaked whale is a deep-water species (500–3,000 m in depth) occurring mainly over continental or island slopes in warm temperate to tropical seas.

IDENTIFICATION

Although reaching lengths of 5.1–6.9 m, it is, however, difficult to observe except in calm seas. Usually it provides only a brief view of its back, and the small, triangular or hooked fin two-thirds of the way along. In some older adults the head and back of the neck are light or almost white, with darker crescent-shaped areas around the eyes. Its head is small, with a sloping or slightly bulbous forehead, short indistinct beak, and up-curved mouth line. A single pair of conical teeth is situated at the tip of the lower jaw, erupting only in adult males and then protruding forward from the mouth. It has small, narrow flippers with a pointed tip, located low down on the flanks and fitting into 'flipper pockets'. Back colouration can be dark, rust brown, slate-grey or fawn. The head and belly are generally paler, particularly in older males. The back and sides are usually covered with linear scars and with white or cream-coloured oval blotches. Sometimes yellow diatoms may occur in patches over the body.

BIBLIOGRAPHY

EVANS, P.G.H., SMEENK, C. & VAN WAEREBEEK, K. 2008. Cuvier's beaked whale *Ziphius cavirostris*. In S. Harris & D.W. Yalden (eds) *Mammals of the British Isles*, 4th edn. Southampton: The Mammal Society. pp. 690–692.

AUTHORS   Peter Evans

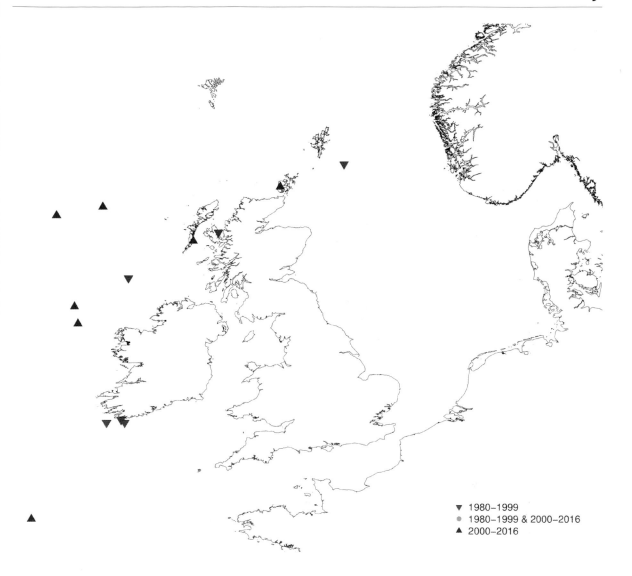

▼ 1980–1999
● 1980–1999 & 2000–2016
▲ 2000–2016

Dwarf sperm whale
Kogia sima (OWEN, 1866)

ROBIN BAIRD

DISTRIBUTION

The dwarf sperm whale has a worldwide, largely tropical, distribution, although its range is known mainly from strandings. In the North Atlantic, it has been recorded from equatorial waters to Virginia (USA) in the west, and to Great Britain in the east. In north-west Europe, the species has been confirmed off the coast of Portugal, where there are also several sightings of unidentified *Kogia* species. Elsewhere, there have been just seven records – five from France, one from Spain and one from the UK. There are also two extralimital records from the Mediterranean, both from Italy. The UK record was of a live animal that came ashore in Cornwall in October 2011, but was successfully refloated.

ECOLOGY

This deep-water species appears to be more restricted to tropical and subtropical seas than the pygmy sperm whale.

IDENTIFICATION

The dwarf sperm whale was not recognised as a distinct species until the mid-1960s and our knowledge of it is rudimentary. It is very similar in appearance to the pygmy sperm whale, having a triangular or squarish head, a narrow underslung lower jaw, and a small robust body that tapers rapidly behind the dorsal fin. As with the pygmy sperm whale, the head shape and light-coloured false gill slit give it a rather shark-like appearance. It is smaller than the pygmy sperm whale, with adults generally reaching lengths of just 2.5 m. The dorsal fin is proportionately larger and more erect – typically 9–16% of the snout to fin length – and located nearer the middle of the back, giving an overall appearance more like a dolphin. The blowhole is also positioned further forward than in its close relative, generally less than 10% of the distance from the tip of the snout. There is usually a pair of short throat grooves similar to those in beaked whales. The flippers are small with bluntish tips, and are situated close to the head. It is brownish-grey in colour over the back and flanks, with a white or pinkish belly.

BIBLIOGRAPHY

McALPINE, D.F. 2017. Pygmy and dwarf sperm whales *Kogia breviceps* and *K. sima*. In B. Würsig, J.G.M. Thewissen & K.M. Kovacs (eds) *Encyclopedia of Marine Mammals*, 3rd edn. London: Academic Press. pp. 786–788.

AUTHORS   Peter Evans

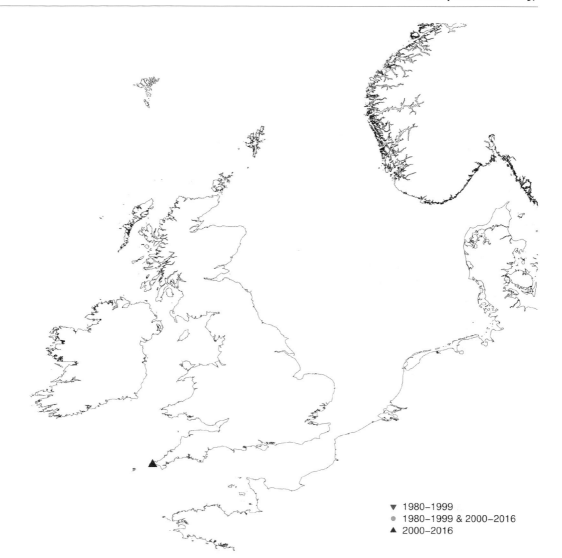

▼ 1980–1999
● 1980–1999 & 2000–2016
▲ 2000–2016

False killer whale

Pseudorca crassidens (OWEN, 1846)

CHRISTOPHER SWANN

DISTRIBUTION

The false killer whale has a worldwide distribution, occurring mainly offshore in deep, warm waters. In the North Atlantic, it occurs from the Equator north to Maryland (USA) in the west, and Norway, Britain and Ireland in the east. In Europe, the species occurs only occasionally north of the Bay of Biscay, although records are largely confined to a few mass strandings from the UK (in 1927, 1934 and 1935), and a handful of sightings from the South West Approaches to the English Channel, west of Ireland, the Hebrides and east of Orkney, all between June and November.

ECOLOGY

The species inhabits mainly tropical to warm temperate seas, greater than 200 m in depth.

IDENTIFICATION

The false killer whale has a long slender body, 5–6 m in length. It is almost all-black in colour except for a blaze of grey on the chest between the flippers, and a grey area that is sometimes present on the sides of the head. The head is small and narrow, and tapers to overhang the lower jaw unlike the similar pygmy killer *Feresa attenuata* and melon-headed *Peponocephala electra* whales. The flippers are long, narrow and tapered, with a distinctive broad hump on the front margin near the middle of the flipper. The dorsal fin, situated slightly behind the middle of the back, is tall and recurved, although its shape can vary from rounded to sharply pointed at the tip. The false killer whale is a fast active swimmer often forming large herds and frequently approaches vessels to bow-ride although, living in deep waters mainly far from the coast, it is only occasionally encountered.

BIBLIOGRAPHY

BORAN, J.R. & EVANS, P.G.H. 2008. False killer whale *Pseudorca crassidens*. In S. Harris & D.W. Yalden (eds) *Mammals of the British Isles*, 4th edn. Southampton: The Mammal Society. pp. 738–740.

AUTHORS Peter Evans

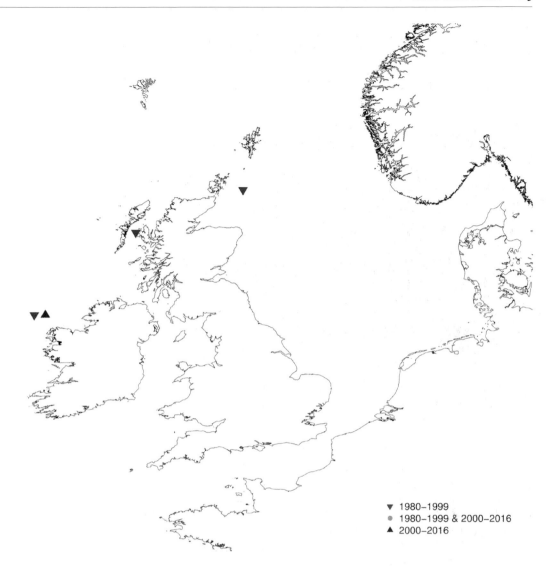

▼ 1980–1999
● 1980–1999 & 2000–2016
▲ 2000–2016

Long-finned pilot whale

Globicephala melas (TRAILL, 1809)

MASSIMILIANO ROSSO, CIMA RESEARCH FOUNDATION

DISTRIBUTION

The long-finned pilot whale is widespread in temperate regions of the world, occurring in the seas of the North Atlantic and Mediterranean, particularly in deep waters off the continental shelf, where it can form groups numbering hundreds of animals.

In the north-east Atlantic, the species is common and widely distributed from the Faroe Islands and Iceland south to the Bay of Biscay and Iberian Peninsula; it is also common in the Mediterranean. In north-west Europe, the main areas that appear to be favoured include the Faroe–Shetland Channel, Rockall Trough, Porcupine Bight, the South West Approaches to the English Channel, and the northern part of the Bay of Biscay. The species is rare in the North Sea except in the northernmost sector, but regularly enters the English Channel. The species occurs in the waters around Britain and Ireland in all months of the year, but with a strong peak of records in July.

ECOLOGY

The long-finned pilot whale usually occurs in deep temperate and subpolar waters of 200–3,000 m in depth (particularly around the 1,000 m isobath), seaward and along edges of the continental shelf where bottom relief is greatest. It may venture occasionally into coastal waters, entering fjords and bays. This behaviour was exploited by humans operating drive fisheries in the Northern Isles and Hebrides up to the early twentieth century, continuing in the Faroe Islands to the present time.

IDENTIFICATION

The species has a robust body with long, pointed, sickle-shaped flippers. Adult males can reach 5–5.9 m and females 3.8–4.8 m. The head is distinctive in shape, being rather square and bulbous, particularly in old males, and with a slightly protruding upper lip. Both the head and back are black to dark grey in colour but an anchor-shaped patch of greyish white can be seen on the chin, which is lighter in younger individuals. The dorsal fin is fairly low, long-based and situated slightly forwards of the midpoint of the back, varying from recurved in immature animals and adult females to flag-shaped in adult males. There is a thick keel on the tailstock, which is more pronounced on adult males. The tail flukes have a concave trailing edge, and are deeply notched in the centre.

BIBLIOGRAPHY

BORAN, J.R., EVANS, P.G.H. & MARTIN, A.R. 2008. Long-finned pilot whale *Globicephala melas*. In S. Harris & D.W. Yalden (eds) *Mammals of the British Isles*, 4th edn. Southampton: The Mammal Society. pp. 735–738.

AUTHORS   Peter Evans

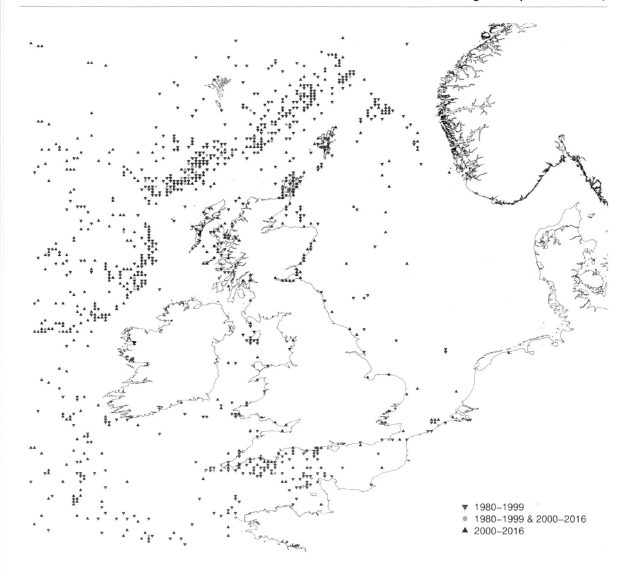

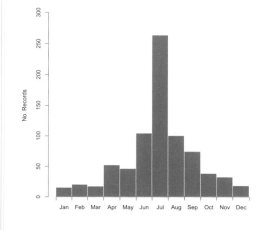

White-beaked dolphin
Lagenorhynchus albirostris (GRAY, 1846)

CHIARA GIULIA BERTULLI, SEA WATCH FOUNDATION

DISTRIBUTION

The white-beaked dolphin is confined to the North Atlantic, occurring from south-west and central east Greenland, Svalbard and the Barents Sea, south to around Cape Cod (USA) in the west and the Bay of Biscay in the east. It is occasionally sighted around the Iberian Peninsula. It is the most common dolphin inhabiting the cold temperate and low Arctic shelf waters of the North Atlantic and North Sea. Four principal centres of high density can be identified: the Labrador Shelf including south-west Greenland; Icelandic waters; the waters around Scotland and north-east England, including the central and northern North Sea and north-west coast of Scotland; and the narrow shelf along the Norwegian coast, extending north into the Barents and White seas. In the UK, the species is most abundant in the northern Hebrides and central and north-western North Sea, occurring also in the southern North Sea, while a small population inhabits the waters of south Devon and Cornwall. Although recorded in all months of the year, most sightings are between June and September.

ECOLOGY

The species occurs in temperate and subpolar seas (generally in temperatures of 2–13°C) of the North Atlantic, including a large part of the north-west European continental shelf, mainly in waters of 50–100 m in depth, and almost entirely within the 200 m isobath. In west Greenland, it occurs in much deeper waters of 300–1,000 m, and in the Barents Sea commonly at 150–200 m and 400 m depths.

IDENTIFICATION

It is a much larger, stouter species than the striped *Stenella coeruleoalba* or common *Delphinus delphis* dolphins, being 2.4–2.8 m in length, with a centrally placed, taller and more recurved dorsal fin. It has a rounded snout and short thick beak that is generally tipped light grey or white. There can often be areas of white also on the head, and generally a black thoracic patch, surrounded by lighter areas. Most characteristic is the pale grey or white area extending along the flanks and over the otherwise dark grey or black dorsal surface behind the fin. The white on the flanks can lead to confusion with the Atlantic white-sided dolphin, although the latter never has white over the back behind the dorsal fin. Juveniles have more uniform colouration, the white areas being more indistinct. The thick tailstock gradually tapers towards the slightly notched tail flukes.

BIBLIOGRAPHY

EVANS, P.G.H. & SMEENK, C. 2008. White-beaked dolphin *Lagenorhynchus albirostris*. In S. Harris & D.W. Yalden (eds) *Mammals of the British Isles*, 4th edn. Southampton: The Mammal Society. pp. 724–727.

GALATIUS, A. & KINZE, C.C. 2016. *Lagenorhynchus albirostris* (Cetacea: Delphinidae). *Mammalian Species* 48 (933): 35–47.

AUTHORS   Peter Evans

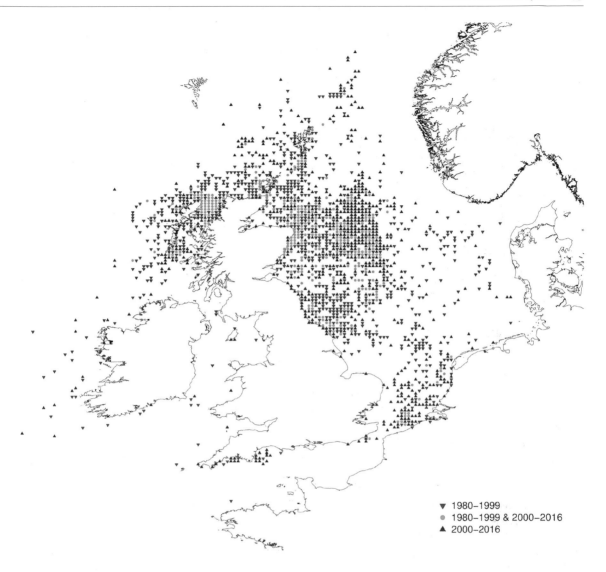

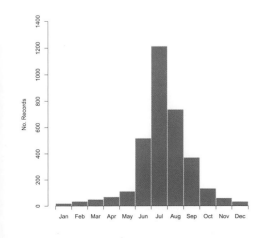

Common dolphin
Delphinus delphis (LINNAEUS, 1758)

KATRIN LOHRENGEL, SEA WATCH FOUNDATION

DISTRIBUTION

The common dolphin has a worldwide distribution in oceanic and shelf-edge waters of tropical, subtropical and temperate seas, occurring in both hemispheres. It is abundant and widely distributed in the eastern North Atlantic, mainly occurring in deeper waters from the Iberian Peninsula north to approximately 65°N latitude (though rare north of 62°N), west of Norway and the Faroe Islands. It occurs westwards at least to the Mid-Atlantic Ridge, and eastwards it enters the western Mediterranean, with a distinct isolated population in the Black Sea. Sightings are rare in the eastern English Channel and the southern North Sea, but are abundant in the Bay of Biscay. On the UK continental shelf, the species is common in the western English Channel and the southern Irish Sea, and further north in the Sea of Hebrides and southern part of the Minch. It is also common south and west of Ireland. In some years, the species occurs further north and east in shelf seas – in the northern Hebrides, around Shetland and Orkney, and in the northern North Sea. Since the 1990s, the species has become regular in the North Sea and even entered the Baltic. Most sightings in coastal waters are between June and October.

ECOLOGY

In the offshore North Atlantic it seems to favour waters over 15°C and shelf-edge features at depths of 400–1,000 m between 49° and 55°N, especially between 20° and 30°W.

IDENTIFICATION

The common dolphin is among the smallest of the true dolphins, being around 1.7–2 m in length. It is an active, fast-moving species, frequently bow-riding boats and jumping clear of the water. The common dolphin has a slender body with a falcate to erect, centrally placed, dorsal fin. In temperate waters, it is the only dolphin species likely to be seen with a long, narrow beak – dark in colour, though sometimes tipped white. On its flanks, there is an hourglass pattern of tan or yellowish tan becoming pale grey behind the dorsal fin. This pale patch may reach the dorsal surface. Underwater, the hourglass pattern shows up very clearly, as does the creamy white belly, contrasting with the brownish black back and upper flanks.

BIBLIOGRAPHY

MURPHY, S., EVANS, P.G.H. & COLLET, A. 2008. Common dolphin *Delphinus delphis*. In S. Harris & D.W. Yalden (eds) *Mammals of the British Isles*, 4th edn. Southampton: The Mammal Society. pp. 719–724.

MURPHY, S., PINN, E.H. & JEPSON, P.D. 2013. The short-beaked common dolphin (*Delphinus delphis*) in the north-eastern Atlantic: distribution, ecology, management and conservation status. *Oceanography and Marine Biology: An Annual Review* 51: 193–280.

AUTHORS   Peter Evans

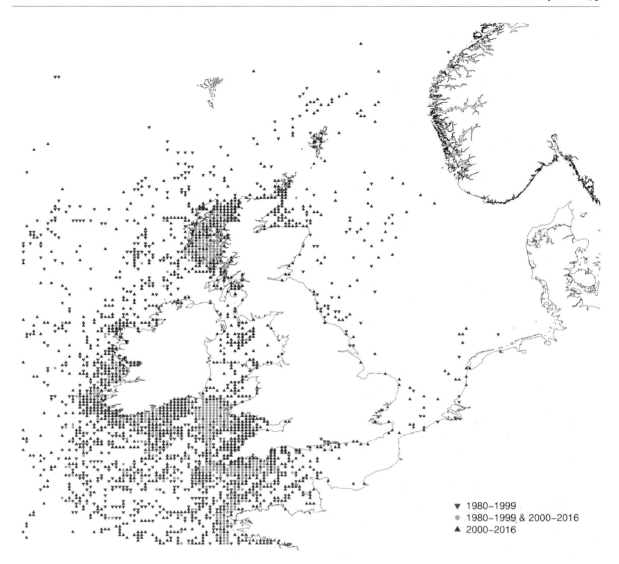

▼ 1980–1999
● 1980–1999 & 2000–2016
▲ 2000–2016

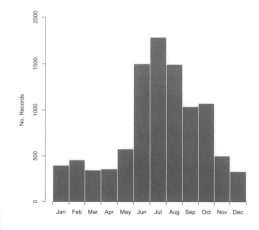

Striped dolphin
Stenella coeruleoalba (MEYEN, 1833)

FRAZER COOMBER, CIMA RESEARCH FOUNDATION

DISTRIBUTION

The striped dolphin has a worldwide distribution, occurring mainly in tropical and warm temperate seas. The species is the most common and widely distributed delphinid in the Mediterranean. In the eastern North Atlantic, it generally occurs further offshore than the common dolphin, with highest densities in the deep waters of the western Bay of Biscay beyond the continental shelf of Spain, Portugal and France. Further north, around Great Britain and Ireland, it is an occasional visitor, recorded mainly from the south-west. However, there has been a sharp increase in records in this region since the 1980s, with the species occurring also in the northern North Sea and even straying into the western Baltic, possibly reflecting warming sea temperatures in the region. The striped dolphin occurs in the mid-Atlantic to at least 62°N, suggesting that its distribution offshore may be extended northward by the Gulf Stream. The species has also been recorded recently from Icelandic, Danish, Swedish and Norwegian waters (with sightings up to 66.5°N). Sightings in UK and Irish waters have all been between May and November, with most in July and August.

ECOLOGY

An oceanic species of warm and temperate waters (mainly 12.5–19.0°C sea surface temperatures), usually occurring well beyond the continental shelf in depths of more than 1,000 m, although it will occasionally come onto the shelf, where it can be recorded in waters of 60 m or less in depth.

IDENTIFICATION

Superficially the striped dolphin resembles the common dolphin, but with a more robust body and a shorter beak (to *c.* 10–12 cm). It lacks yellow patches and has two distinct black stripes on the flanks, one from the eye to the flipper, and the other to the anus. A distinct groove separates the beak from the forehead. Also evident is a white or light grey V-shaped blaze, one branch of this narrowing to a point below the dorsal fin, the other extending backwards towards the tail. Adults are between 1.9 and 2.4 m in length. The colouration of the body is variable, dark grey or bluish grey on the back, with lighter grey flanks, the posterior part of which is light grey and sometimes extends upwards over the dorsal surface of the tailstock. The belly is white. The dorsal fin is slender, recurved and centrally placed. The tailstock is narrow with no obvious keel, and the dark tail flukes have a median notch in them.

BIBLIOGRAPHY

EVANS, P.G.H. & COLLET, A. 2008. Striped dolphin *Stenella coeruleoalba*. In S. Harris & D.W. Yalden (eds) *Mammals of the British Isles*, 4th edn. Southampton: The Mammal Society. pp. 715–719.

AUTHORS   Peter Evans

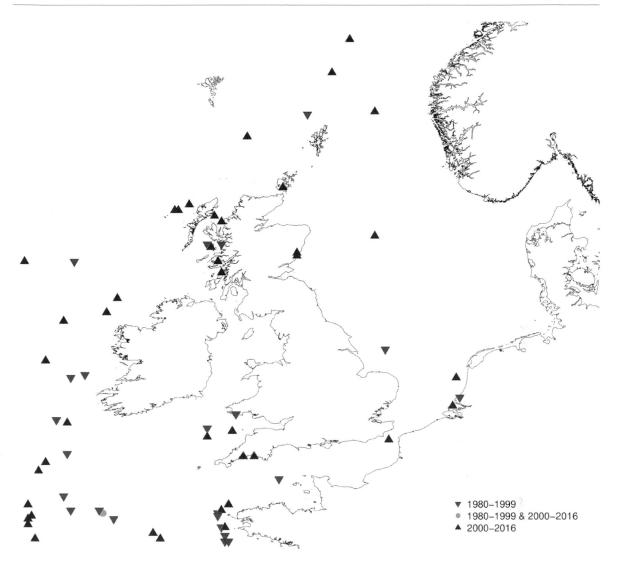

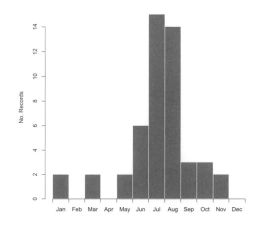

Bottlenose dolphin

Tursiops truncatus (MONTAGU, 1821)

DAVE POWELL, SEA WATCH FOUNDATION

DISTRIBUTION

The bottlenose dolphin has a worldwide distribution in tropical and temperate seas in both hemispheres. In the North Atlantic, it occurs mainly from Nova Scotia in the west and the Faroe Islands in the east, southwards to the Equator and beyond. Along the Atlantic seaboard of Europe, it is locally fairly common near-shore off the coasts of Spain, Portugal, north-west France, western Ireland (particularly the Shannon Estuary and Connemara), north-east Scotland (particularly the Moray Firth south to the Firth of Forth), south-west Scotland, in the Irish Sea (particularly north and west Wales, including all of Cardigan Bay), and in the English Channel (particularly around Cornwall and the Channel Islands/Normandy coast). Smaller groups have also taken up residence at other localities – for example, around the Outer Hebridean island of Barra, and in the Inner Hebrides (Islay, Mull, Coll, Tiree and southern Skye) in west Scotland. Although present all year round, most sightings are between May and September.

The species also occurs offshore in the eastern North Atlantic, especially along the shelf edge, as far north as the Faroe Islands and even Svalbard. In the Bay of Biscay, there are particularly high numbers over the outer shelf and shelf break. During the summer, some pelagic groups may enter near-shore waters around the Faroe Islands, northern and western Scotland, western Ireland, in the Bay of Biscay, and around the Iberian Peninsula.

ECOLOGY

In coastal waters, the bottlenose dolphin often favours river estuaries, headlands or sandbanks where there is uneven bottom relief and/or strong tidal currents. Offshore, the species tends to range along the shelf edge.

IDENTIFICATION

A relatively large, stout dolphin, the bottlenose dolphin reaches adult lengths of 3–3.8 m. Unlike many other dolphin species, it lacks distinctive markings, being dark grey on the back, lighter grey on the flanks, and grading to white on the belly. When breaching backwards, it displays its white throat and belly and bottle-shaped nose. The rounded head has a distinct short beak, often tipped with white on the lower jaw. The flippers are fairly long and pointed and the centrally placed dorsal fin is tall, slender and recurved, although its size and shape can be very variable. Nicks along the edge of the dorsal fin, and scratches on the fin and back, are commonly used to identify individuals.

BIBLIOGRAPHY

WILSON, B. 2008. Bottlenose dolphin *Tursiops truncatus*. In S. Harris & D.W. Yalden (eds) *Mammals of the British Isles*, 4th edn. Southampton: The Mammal Society. pp. 709–715.

AUTHORS   Peter Evans

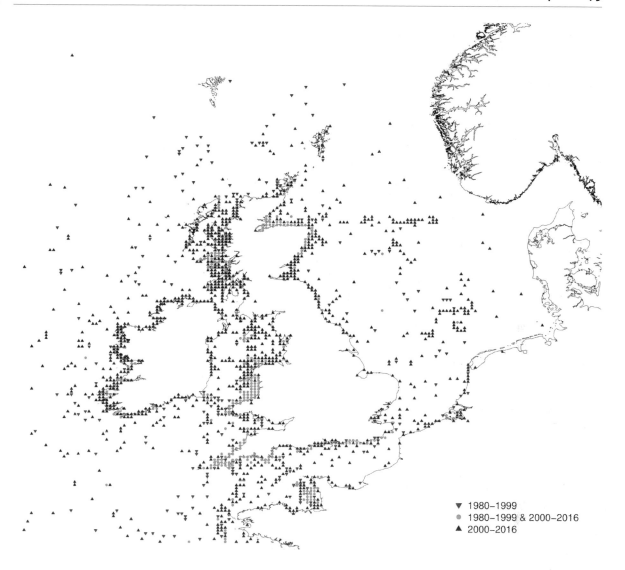

▼ 1980–1999
● 1980–1999 & 2000–2016
▲ 2000–2016

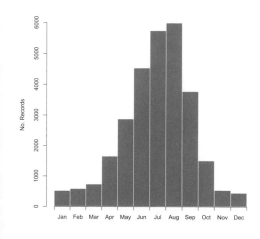

Harbour porpoise
Phocoena phocoena (LINNAEUS, 1758)

PETER EVANS, SEA WATCH FOUNDATION

DISTRIBUTION

The harbour porpoise is confined to the North Atlantic and North Pacific. In the North Atlantic, the species occurs mainly from central west Greenland and Novaya Zemlya in the north to North Carolina and Senegal in the south. There is a geographically isolated population in the Black Sea. It is the commonest and most widely distributed species of cetacean in northern European seas, occurring over the continental shelf from the Barents Sea and Iceland, south to the coasts of France and Spain. In the 1970s it became scarce in the southernmost North Sea, English Channel and Bay of Biscay, but the species has returned to the southernmost North Sea, English Channel and French Biscay coast since the 1990s. There is some evidence for a southward shift from the north-western to the south-western North Sea. The harbour porpoise occurs all year round in UK and Irish waters, although most sightings are made between June and September.

ECOLOGY

The species is restricted to temperate and subarctic (mainly 11–14°C) seas, mainly over the continental shelf at depths of 20–200 m. Although the harbour porpoise can be found in deep waters off the edge of the European continental shelf (for example within the Faroe Bank Channel), it is comparatively rare in waters exceeding 200 m in depth. Radio-tagged porpoises from west Greenland have been tracked seasonally to the mid-Atlantic at depths of 1,000–3,000 m. In coastal waters, the species frequently uses tidal conditions for foraging.

IDENTIFICATION

It is the smallest cetacean in European seas, with adults averaging a length of around 1.5 m. It only occasionally leaps clear of the water, and usual views are of a small, centrally placed triangular dorsal fin and a glimpse of the back. It has a small rotund body, and a small head with no forehead or beak. The flippers are short and rounded. Its back is dark grey with a paler patch on the flanks that extends up the sides in front of the dorsal fin. From the flippers, a dark grey line extends to the jaw-line.

BIBLIOGRAPHY

EVANS, P.G.H., LOCKYER, C.H., SMEENK, C., ADDINK, M. & READ, A.J. 2008. Harbour porpoise *Phocoena*. In S. Harris & D.W. Yalden (eds) *Mammals of the British Isles*, 4th edn. Southampton: The Mammal Society. pp. 704–709.

AUTHORS   Peter Evans

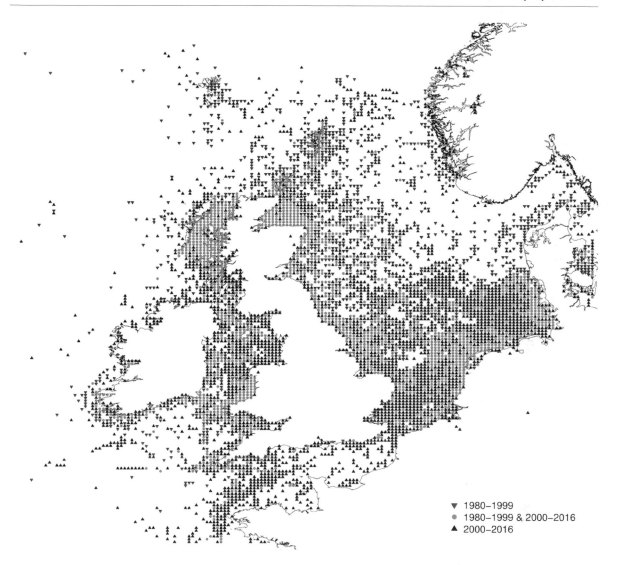

▼ 1980–1999
● 1980–1999 & 2000–2016
▲ 2000–2016

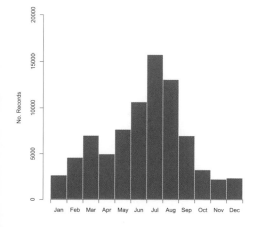

CETACEANS KNOWN IN BRITAIN AND IRELAND ONLY FROM STRANDINGS

ROBIN BAIRD

Blainville's beaked whale
Mesoplodon densirostris (BLAINVILLE, 1817)

In the eastern North Atlantic, the Blainville's beaked whale has been recorded from Spain, Portugal, Madeira and the Canaries, with strandings recorded in the Netherlands, Iceland, and France. There are no live sightings from UK or Irish waters. However, there has been a single extralimital stranding record from Wales in July 1993.

The Blainville's beaked whale is a deep-water species inhabiting ocean basins and trenches of 500–5,000 m in depth. It is a medium-sized beaked whale (4–5 m) with a body shape typical of the genus, which is bluish grey or black in colour and often marked with pale ovals and linear scars. This species has a strongly arched jaw, and in males a massive pair of flattened triangular teeth erupt at the apex (often encrusted with barnacles).

GONZALO MUCIENTES SANDOVAL

Gervais' beaked whale
Mesoplodon europaeus (GERVAIS, 1855)

The Gervais' beaked whale is only known from the Atlantic and recently several strandings have been recorded in the eastern North Atlantic. These include western Ireland (January 1989), France, Portugal, southern Spain, the Azores, Mauritania and Guinea-Bissau, although most records have come from the Canaries. Except for the type specimen found floating in the English Channel in 1848, there are no sightings or strandings from UK waters.

This species is a medium sized (4.5–5 m) beaked whale, with a dark grey or indigo back that becomes light grey on the lower flanks and belly. It is distinguished from other beaked whales by the short, but clearly defined, slender beak, and relatively straight mouth line. In adult males one pair of laterally compressed, triangular teeth erupt in the lower jaw about one-third of the way along the gape.

BRIAN CLASPER, ORCA

True's beaked whale
Mesoplodon mirus (TRUE, 1913)

Since 1899, there have been 11 records of stranded True's beaked whale, ten of which were from the west of Ireland. There are no definite records from UK waters but sightings based on appearance in the eastern North Atlantic (not all confirmed by genetics) have been made in the Canaries, Azores and Bay of Biscay.

The most diagnostic feature of this species is a short but clearly defined beak, sloping into a slightly bulbous forehead. There is a single pair of slightly laterally compressed teeth, oval in cross section, directed forward and upward at the extreme tip of the lower jaw, exposed above the mouth line only in adult males. The body is typical of the genus (4–5.5 m), slate grey in colour and often accompanied by pale spots and linear scars. A dark eye patch may also be present.

SHUTTERSTOCK, WILDESTANIMAL

Narwhal
Monodon monoceros (LINNAEUS, 1758)

The narwhal rarely occurs outside of the artic, however, since the sixteenth century there have been seven records from Britain and Ireland. The only and last sighting of a narwhal in UK waters was of two individuals off Orkney in June 1949. Elsewhere in western Europe, the only records are from Germany (1736), the Netherlands (1921), Sweden (1992) and Belgium (2016).

The narwhal is a distinctive 5 m-long cetacean (excluding tusk). Its body is stout with a small rounded head with bulbous forehead and very slight beak. It is mottled grey-green, cream and black in colour but older males appear lighter, partly because of the accumulation of white scar tissue. The left tooth of the male narwhal is greatly extended (up to 2.7 m long) in all but the youngest animals and erupts through the upper lip as a spiral tusk.

ROBIN BAIRD

Melon-headed whale
Peponocephala electra (GRAY, 1846)

The only records of the melon-headed whale in north-west Europe are from a stranding in Charlestown, Cornwall, in September 1949, and from three individuals that live-stranded near La Rochelle in France (two in 2003, and another in 2008). This species is generally found in waters deeper than 1,000 m.

The species reaches 2.3–2.7 m in length, with a torpedo-shaped body that is uniformly dark in colour but often has white lips and a subtle grey cape pattern over the back. The dorsal fin is relatively tall, recurved and centrally placed on an arched back.

ROBIN BAIRD

Fraser's dolphin
Lagenodelphis hosei (FRASER, 1956)

In the eastern North Atlantic Fraser's dolphin occurs mainly off the coast of north-west Africa and in the Canaries, the Azores and Madeira. However, there was a mass stranding of 11 animals in north Brittany on the Atlantic French coast in 1984, and a single stranding record from the Outer Hebrides, north-west Scotland, in 1996.

With characteristics of both the common and white-beaked dolphin, it has a robust body (2.4–2.7 m in length) with a short but well-defined beak and small curved dorsal fin and flippers. The body is bluish grey on the back and light pink or white on the belly. A creamy white band and parallel black band extend from the eye along the flanks towards the anus.

BIBLIOGRAPHY

EVANS, P.G.H. 2008. Fraser's dolphin *Lagenodelphis hosei*. In S. Harris & D.W. Yalden (eds) *Mammals of the British Isles*, 4th edn. Southampton: The Mammal Society. pp. 731–733.

EVANS, P.G.H. 2008. Melon-headed whale *Peponocephala electra*. In S. Harris & D.W. Yalden (eds) *Mammals of the British Isles*, 4th edn. Southampton: The Mammal Society. pp. 733–735.

EVANS, P.G.H., AGUILAR DE SOTO, N., HERMAN, J.S. & KITCHENER, A.C. 2008. Blainville's beaked whale *Mesoplodon densirostris*. In S. Harris & D.W. Yalden (eds) *Mammals of the British Isles*, 4th edn. Southampton: The Mammal Society. pp. 697–699.

EVANS, P.G.H., HERMAN, J.S. & KITCHENER, A.C. 2008. Gervais' beaked whale *Mesoplodon europaeus*. In S. Harris & D.W. Yalden (eds) *Mammals of the British Isles*, 4th edn. Southampton: The Mammal Society. pp. 696–697.

EVANS, P.G.H., HERMAN, J.S. & KITCHENER, A.C. 2008. True's beaked whale *Mesoplodon mirus*. In S. Harris & D.W. Yalden (eds) *Mammals of the British Isles*, 4th edn. Southampton: The Mammal Society. pp. 694–696.

MARTIN, A.R. & EVANS, P.G.H. 2008. Narwhal *Monodon monoceros*. In S. Harris & D.W. Yalden (eds) *Mammals of the British Isles*, 4th edn. Southampton: The Mammal Society. pp. 702–704.

VAGRANT SPECIES AND THOSE WITHOUT ESTABLISHED POPULATIONS IN THE UK

MARK BALDWIN

Raccoon
Procyon lotor (LINNAEUS, 1758)

The raccoon is a cat-sized carnivore with grey fur, a distinctive black band across its eyes, and a series of black bands on its tail. It has a plump body with a hunched back and pointed nose and ears. This species has been spotted in a few locations and these records probably relate to escaped individuals. No breeding populations are known and there are no presence records of this species in the databases used to create this Atlas. However, the high potential for this species to become invasive means that ongoing surveillance is essential.

MARK BALDWIN

Red-necked wallaby
Macropus rufogriseus (DESMAREST, 1817)

Between 2000 and 2016 there were 132 verified sightings of the red-necked wallaby from 23 different hectads. This includes 105 records from the Peak District, 11 records from the Isle of Man, 11 records in Buckinghamshire and Bedfordshire, and two in Yorkshire. There have also been individual records from Norfolk, Devon and Loch Lomond.

The red-necked wallaby is the only species of large marsupial known to be present in the UK. Feral colonies of this species began from escaped zoo stock in the 1940s with colonies in the Peak District, the Weald, Sussex, and Inchconnachan Island in Loch Lomond.

EUGENE BUTTERWORTH

Reindeer
Rangifer tarandus (LINNAEUS, 1758)

There are 208 records from seven hectads since the year 2000 of reindeer in the UK. Most of these are from within a few hectads in the Cairngorms National Park. There is also a single record east of Edinburgh.

The reindeer is a domesticated species that was introduced to the UK in 1952 from Sweden (wild reindeer being extinct in the UK since the Mesolithic period) and, despite being free-ranging, the Scottish population is privately owned and managed. The reindeer is a robustly built medium-sized deer, standing around 85–150 cm tall at the shoulders. Both sexes have multi-branched antlers. It has a thick dense coat (thicker on the shoulders) that is greyish brown, and often lighter upper parts.

HENRY SCHOFIELD

Pond bat
Myotis dasycneme (BOIE, 1825)

The pond bat has a distribution stretching from north-western France and southern Scandinavia south to Serbia and Montenegro, Ukraine, and north Kazakhstan, east to central Russia. However, within north-west Europe, records are patchy. There is currently only one record of this species in the UK. That was of an adult male found in Kent in 2004.

The pond bat is a medium sized bat (13–18 g) with dense pale grey-brown dorsal fur and well-demarcated white/light grey ventral fur. The skin is reddish to light brown on the face, whilst the skin elsewhere is grey-brown. The feet are noticeably large and hairy, and the tragus – which is unusually short for a *Myotis* species – has a blunt tip and is bent slightly inwards.

HENRY SCHOFIELD

Geoffroy's bat
Myotis emarginatus (GEOFFROY, 1806)

The Geoffroy's bat is distributed throughout the Mediterranean region. In the UK there are a very small number of records from swarming sites. These include two individuals captured in 2012 and 2013 in West Sussex; and an animal with the morphological characteristics of the species but lacking genetic confirmation, captured in 2013 in Wiltshire.

This medium sized bat (6–9 g) is very similar in appearance the Natterer's bat, so some under-recording is possible. The key characteristics distinguishing it from the Natterer's bat are the long rust-brown coloured woolly fur and a distinctive right angled notch on the outer ear margin. The tragus is long and thin but does not reach the ear notch.

JENS RYDELL

Northern bat
Eptesicus nilssonii (KEYSERLING AND BLASIUS, 1839)

The northern bat is distributed across central and eastern Europe, through to France and Northern Italy in the west, and into the Arctic Circle in the north. It is generally considered to be a sedentary species, though there are occasional records of very long distance movements (>450km). In the UK there are three isolated records for the species that appear to derive from natural movements. These include a hibernation record in Surrey (1986), one on a North Sea oil platform (1996) and one in Berkshire (2014).

The northern bat is similar in appearance to the serotine, with dark brown skin, broadly rounded ears, and a tragus that is broadest in the middle. It is smaller (9–13 g) than the serotine bat. The ventral fur is characteristic, being dark with golden-tips, and is well-demarcated from the yellow-brown ventral fur, particularly around the throat. The ears are broadly rounded, do not meet in the middle and have a tragus that is broadest in the middle.

SALLY-ANN HURRY

Parti-coloured bat
Vespertilio murinus (LINNAEUS, 1758)

The parti-coloured bat is distributed throughout most of northern Europe with a scattering of records from the UK. Since 2000 there have been 15 records from 12 different hectads, six of which have come from the Shetland Isles. The other records are derived from across Britain, including two on the Isle of Wight, four near Eastbourne, one in Oxfordshire and one on the Isle of Arran.

The parti-coloured bat is a medium-sized bat (10–15 g) with short, broad, rounded ears. It has dark brown or black fur that is longer on the back and tipped with silver, giving a frosted appearance. The face and ears are blackish brown in colour.

DANIEL WHITBY

Kuhl's pipistrelle bat
Pipistrellus kuhlii (KUHL, 1817)

Kuhl's pipistrelle is a Mediterranean bat species and is distributed throughout southern Europe and northern Africa. Since 2000, there have been eight records of Kuhl's pipistrelle in the UK. These are mainly derived from south-east England, with four records from the Isle of Wight, one in the Thames Valley, two near Colchester and one in Bourne. The species is also regularly recorded along the coast of parts of south-east England in acoustic surveys. Records of Kuhl's pipistrelle in the UK have been attributed by some as translocations and by others as colonisation from a species with an increasing northward range expansion.

This species is very similar in appearance to other European pipistrelle bats (5–10 g), but is distinguished by its dentition: the incisors have a single cusp whereas all other pipistrelle bats have two cusps. It may therefore be under-recorded.

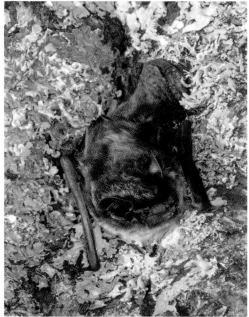

HENRY SCHOFIELD

Savi's pipistrelle bat
Hypsugo savii (BONAPARTE, 1837)

The Savi's pipistrelle has a wide distribution stretching from the Mediterranean and North Africa through to the Middle East and Caucasus to northern India. Over the last 30 years, it has expanded northwards in central Europe, with some individuals being found to migrate to Germany. In the UK, individuals that appear to have arrived naturally have been recorded in Eastbourne (1993) and Merseyside (1996), but there are no records since 2000.

This species has the typical general characteristics of a pipistrelle bat (5–9 g). However the dorsal fur is longer, and the golden-yellowish tip to the brown fur are noticeable. The ears are relatively short with a tragus that broadens at the apex, being much broader than that of a common or soprano pipistrelle. The face and ears are very black, and somewhat shiny. The penis has a right-angled bend near the base, and the tail projects 4–5mm from the tail membrane.

There are five species of phocid or true seals that are normally distributed in the higher latitudes of the Atlantic Ocean but which have occasionally been recorded around the UK.

L. JACKSON

Bearded seal
Erignathus barbatus (ERXLEBEN, 1777)

Since 2000 there have been eight records of the bearded seal from six different hectads. Records are mainly on the Shetland Isles, but two sightings were also made in the Moray Firth and one on the Northumberland coast.

Adults of both sexes are around 2.2–2.7 m in length from nose to tail and weigh around 235–340 kg. The bearded seal has a distinctively small head with characteristic vibrissae, giving the bearded appearance. The colour is generally grey with touches of brown, especially around the head.

SHUTTERSTOCK, TODD BOLAND

Harp seal
Pagophilus groenlandicus (ERXLEBEN, 1777)

Since 2000 there have been ten records of harp seal sightings from six different hectads. Records include six in Cornwall, two in Liverpool Bay, one near Newcastle and one from the Pembrokeshire coast.

Adults of both sexes are up to 1.7 m in length from nose to tail and weigh around 130 kg. It is easily distinguished by its dark face-mask and by the large 'harp'-shaped dark band along each flank starting at the shoulders.

SHUTTERSTOCK, ENRIQUE AGUIRRE

Hooded seal
Cystophora cristata (ERXLEBEN, 1777)

Since 2000, there have been 11 records of hooded seals from ten different hectads. The records include two in Cornwall, two in Devon, two from Norfolk and the Wash, four between Newcastle upon Tyne and Whitby, and one from the Shetland Isles.

Males are larger than females, reaching 2.2–2.5 m in length and weighing over 400 kg compared with 2.2 m and 300 kg for females. Males also have a very obvious inflatable hood ornament. Both sexes are generally grey in colour and have numerous dark patches of irregular sizes across the body.

SHUTTERSTOCK, CHONLASUB WORAVICHAN

Ringed seal
Pusa hispida (SCHREBER, 1775)

Since 2000 there have been two sightings of ringed seals, one in Shetland and one on the Norfolk coast.

Adults from both sexes are around 1.5 m in length from nose to tail and weigh around 45–95 kg. Females are generally smaller than males. This species is very similar in appearance to the common seal. It is grey with black spots that often have pale edges, giving the species its common name.

L. JACKSON

Walrus
Odobenus rosmarus (LINNAEUS, 1758)

Since 2000 there have been two records of walruses in British waters. They both occurred in Orkney in 2013 and in 2018.

The walrus is a large seal species growing up to 3.65 m and weighing in at 1,270 kg for an adult male, females being much smaller at 3 m and 850 kg. This species is very distinctive with unique dentition of large tusks – which are present in both sexes – and its flushed red colour when hauled out.

BIBLIOGRAPHY

BAKER, S.J. & HILLS, D. 2008. Species that have survived for more than 1 year, but not bred. In S. Harris & D.W. Yalden (eds) *Mammals of the British Isles*, 4th edn. Southampton: The Mammal Society. pp. 783–793.

DANSIE, E., PUTMAN, R.J. & YALDEN, D.W. 2008. Reindeer *Rangifer tarandus*. In S. Harris & D.W. Yalden (eds) *Mammals of the British Isles*, 4th edn. Southampton: The Mammal Society. pp. 604–605.

DIETZ, C. & KIEFER, A. 2016. *Bats of Britain and Europe*. London: Bloomsbury.

HALL, A.J. 2008. Vagrant seals. In S. Harris & D.W. Yalden (eds) *Mammals of the British Isles*, 4th edn. Southampton: The Mammal Society. pp. 547–550.

HUTSON, A.M. 2008. Northern bat *Eptesicus nilssonii* In S. Harris & D.W. Yalden (eds) *Mammals of the British Isles*, 4th edn. Southampton: The Mammal Society. pp. 360–361.

HUTSON, A.M. 2008. Pond bat *Myotis dasycneme* In S. Harris & D.W. Yalden (eds) *Mammals of the British Isles*, 4th edn. Southampton: The Mammal Society. p. 323.

HUTSON, A.M. 2008. Savi's pipistrelle bat *Hypsugo savii*. In S. Harris & D.W. Yalden (eds) *Mammals of the British Isles*, 4th edn. Southampton: The Mammal Society. p. 356.

AUTHORS   Fiona Mathews and Frazer Coomber

FERAL COLONIES AND POPULATIONS

There are several feral animals recorded in the UK. However, because of their domestic background there are many caveats that relate to these records. A true feral population is a wild self-sustaining population of escaped or released individuals descended from domesticated stock. A good example of a feral population is the Soay sheep *Ovis aries* that live on the islands of St Kilda. Moreover, the feral species' records that were collated during the Atlas are likely to include a large number of domestic or managed individuals that are impossible to distinguish without additional information.

MARK BALDWIN

Feral ferret
Mustela putorius fero

Since 2000 there have been 610 records from 306 different hectads of the feral ferret, and 460 records from 26 hectads of ferret–polecat hybrids. Almost all counties of Great Britain and Northern Ireland have records of polecat-ferrets and feral ferrets, but the records are most numerous for southern England and the east Midlands. There are also records from many of the offshore islands including the Isle of Wight, Anglesey, the Isle of Man, Arran, Jura, Mull, Eigg, Skye, the Uists and Benbecula, Shetland and Rathlin.

JAMES MILLER

Feral goat
Capra aegagrus hircus

Since 2000 there have been 515 records of the feral goat across the UK from 89 different hectads, with notable record clumps in Gwynedd, Wigtown, Stirling and Falkirk, Inverness, and Ross and Cromarty. The feral goat is also present on a number of islands including the Isle of Wight, Lundy, Islay, Jura, Kerrera, Seil, Mull, Rum and Skye.

VIVI BOLIN

Feral sheep
Ovis aries

Since 2000 there have been 42 records of the feral sheep from 27 different hectads. These are mainly from Britain, with noticeable groupings of records in the Cairngorms, Cumbria and South Yorkshire. There are also records from Lundy and the Isle of Wight.